AF567058

CHRONICLES
of a
CONCERNED
CONSULTANT

CHRONICLES
of a
CONCERNED
CONSULTANT

Eugene W. Greenfield

Illustrated by
Eugene and Eric Greenfield
and
John Goddard

VANTAGE PRESS
New York / Los Angeles / Chicago

FIRST EDITION

Published by Vantage Press, Inc.
516 West 34th Street, New York, New York 10001

Manufactured in the United States of America
ISBN: 0-533-08263-3

Library of Congress Catalog Card No.: 88-90447

To engineers and all
discerning people everywhere

Contents

Preface ix

The Case of the Deadly Back-feed 1
The Case of the Burning Tunnel 7
The Case of the Poor Painter 14
The Case of the Unbaked Bread 20
The Case of the Foolish Fireman 25
The Case of the Rough Plowman 30
The Case of the Big Intake 35
The Case of the Lost Sunday Dinners 40
A Case of Mistaken Identity 45
The Case of the Twisted Bars 49
The Case of the Falling Fuse Clip 54
The Case of the Distressed Professionals 61
The Case When the Off Was On 66
The Case of the Dam Dilemma 76
The Case of the Dreadful Dredge 83
The Case of the Hidden Casualty 89
The Case of the Disappearing Armor 93
The Case of the Park's Power Problem 98
The Case of the Shocking Landlord 104
The Case of the Mysterious Mines 110
The Case of the Hotel Air Conditioners 116
The Case of the Forbidden Climb 124
The Case of the Forgetful Supervisors 129
The Case of the Unhelpful Consultant 133
The Case of the Hanging Hazard 137

The Case of the Steaming Wells 140
The Case of the Deadly Bight 148
The Case of the Wooden Lifesaver 154
The Case of the Worried Mayor 156
The Case of the Sleepy Physician 159
The Case of the Boss's Romance 165
The Case of the House without a Ground 171
The Case of the Double-set Poles 176
The Case of the Ancient and Honorable 182
The Case of the Crossed Connections 188
The Case of the Judge Who Said No 195
The Case of the Lineman's Helper 205
The Case of the Frightened Executives 211
The Case of the Untested Switch Box 215
The Case of the Ocean Energy Gamble 222
The Case of the Hawaiian Hiatus 230

Preface

Almost anyone after a lifetime of professional practice can look back at the many events or happenings of his career that were particularly interesting, humorous, emotionally charged, or even bizarre—with strange and unforeseen endings. Some may have turned out to be triumphant successes; others were miserable failures. But on the whole, now that I am safely retired I can say it was all a very satisfying experience and out of this kaleidescope of years those events that have particularly stayed with me still arouse the old intense interest.

My career, electrical power engineering, spanning almost sixty years, especially brought me into situations of the most varied kinds—with industry, with educational institutions, with governments, and latterly with the legal profession investigating electrical accidents, all of which leads me to answer your natural query, "Why this book?"

I think there are several reasons for this book. Of course one might say that I want to document my "great" contributions to society. No, I haven't a shred of that object or feeling. Or I may be wanting to relive my active experiences in print. Yes, it's partly that, I am sure. However, the overriding reason for going to all the (delightful) trouble of writing and getting published is that you, the reader, will, and I think you should, become very interested, perhaps fascinated, by what can and does happen to a concerned consultant in the electrical pursuits of this modern world.

Of course the background of each of the events or cases you will read about is technical. I hope, however, I have been able to popularize the technical concepts so well that a nontechnical background should in no way prevent you from understanding the development of the story and its outcome. In that rare event that there might be something you cannot completely grasp, just write me, care of the publisher.

During my career span as indicated above, I have worked as an employee and consultant for and in a great many areas. For example, my notes show 225 separate consulting engagements. For this book I have selected some forty-one of these, and they well highlight the humorous, the emotional, the strange, and my good and not so good contributions to clients and the profession.

Obviously there are literally hundreds of people—industry executives, managers, engineers, lawyers, and just plain folk—who in some way contributed to the career I so glibly called my own. To all of these—and since it is impossible to recall each one, most will remain anonymous—I give my sincere thanks.

The reader will note the many areas where I worked with individual lawyers and law firms. My years of experience as an expert for the attorneys representing my clients and also with our opposing attorneys were very satisfying and enlightening. I found these attorneys knowledgeable, hardworking, highly ethical, and courteous. I can recall only one instance when an opposing lawyer tried to browbeat me during a deposition, and that was quickly squashed by our attorney.

Another thing to say. Since this book describes real, not fictional, events and these were agencies, companies, and real people and actual places involved, for obvious reasons I have omitted all but a few such names and in most cases used fictional designations and unidentified places.

I wish to express warm appreciation to my son Eric for his helpful criticism of language, sharp at times but effective, and for the most part heeded, also for his assistance with the many line illustrations, about half of which he and I prepared. The balance were elegantly done by Mr. John Goddard of Fallbrook.

And finally there is that person to whom I give sincerest thanks and gratitude for her warm encouragement, her wonderful patience, and her expert understanding secretarial work, not only on the manuscript for this book but also on our many good trips together to unravel the problems of our clients. Yes, she is my wife, Louise, and as we near our sixtieth anniversary I know no man was better blessed.

CHRONICLES of a CONCERNED CONSULTANT

The Case of the Deadly Back-feed

Oregon is a beautiful state bounded by majestic mountains, broad rivers, a desert, and the ocean. Its central areas are lush with greenery and flowers, its mountains reaches with fantastic frozen lava floes. As big cities go, Oregon has very few, but they and the many small communities are notable for their floral beauty. Where most Oregonians dwell the climate is generally temperate and living is easy.

There are occasions, however, when Nature turns its smiling face and unleashes a concentrated violence on an unsuspecting and unprepared populace. During the winter months perhaps once a decade there will be an ice storm, covering everything with an inch or more of shaped ice, snapping trees and power lines and stopping all driving and even walking activity. Toward late fall, though this is still a rare occurrence, winds of up to one hundred miles an hour can slam into a community and batter and tear loose everything for two or three days.

That is what happened sometime in mid-November a few years ago when a major city was struck by such a wild storm.

Property destruction was extremely heavy in some localities and comparatively light in others. Where the winds struck most heavily the well-treed residential areas took the worst beating. Big tree limbs and even whole trees came down and took along many of the overhead primary and secondary power lines, so that a large percentage of the electric power was lost in the affected areas and many brownouts occurred in adjacent streets.

Early in the first day of the storm, as multitudes of calls came into the local power utility telling of lost power, dispatchers started bringing in all available line, service, and maintenance crews for emergency protection of the public and securing of downed lines in the streets and yards. Clean-up would have to wait for the storm's abatement.

Sometime during the early morning of Sunday, a forty-year-old building contractor decided to check on some remodeling he had started on an old mansion in a southwest suburb of the city. He had heard radio reports of severe damage in that area. The old mansion was set back some two hundred yards from A Street, and there was an access drive and gate (usually open) at the street entrance. As he turned off that suburb's main thoroughfare, D Boulevard, into A Street, he had to proceed slowly due to many tree limbs and brush debris on the roadway. It had been raining in torrents only an hour before but now had slackened off. The ditches on both sides of the street were overflowing, and there was standing water over the street itself.

As the contractor approached the mansion driveway, he saw that one of the overhead power conductors had broken down and was lying across the entrance. This was one of four wires on a 13,800-volt primary pole line running along A Street from a tap off the main service feeder line on D Boulevard. As the accident was later reconstructed, he got out of his pickup and with a piece of broken tree limb gingerly pushed the downed wire back out of the way. Although the wire was partly lying in water, there were no sparkings or smoking from it, and by that he must have concluded that it was not energized. (And he was right about that.) So he pushed it completely out of the way and went on to open the gate, which had been slammed shut by the wind.

Immediately in front of the gate, the other end of the broken line wire was also lying on the ground. Being a building contractor familiar with electric wiring as such, the man had logically reasoned that since the wire he had just moved with the stick came from the source feeder on D Boulevard and was dead, the other part of the broken line went to its consumer loads and therefore was dead also. So he reached down and picked up the wire to move it. He was electrocuted instantly. The supposed dead line was at many thousand volts to ground!

I will not here go into the manner of this man being found by a passing motorist and a young boy taking a shortcut across the mansion field or the attempts at resuscitation or how his probable actions leading up to the fatal shock were reconstructed. What was still to be determined was how it was possible for the load or customer end of the broken-down line wire to have remained energized when the source end of the same wire from which the power came was plainly dead—also what and who was responsible for that situation.

The attorney representing the contractor's family called me into the case to help him determine what had happened and how it happened. I visited the site several times and charted all the pole lines and every connection from these to the consumer loads around the area where the accident had occurred. I found that the tap-off from the main feeder line on D Boulevard was a three-phase line with a neutral wire; that is, it consisted of four wires, three of which were energized, and the neutral wire was grounded. Voltage between pairs of the energized conductors was 13,800, and between any one of these and ground it was 7,900 volts.

Since the houses on A Street and adjacent streets were all served from single phase transformers—meaning only one energized conductor and the neutral were all that was required—I was surprised to see that the tap-off was a three-phase circuit. Why would a suburban residential area need three-phase power lines? The answer to this was soon found. The area was not all residential. Just a few blocks to the north on a street intersecting A Street was a large junior high school. Due to its heavier electrical power requirements, type of motors, et cetera, it was supplied through three-phase transformers and so required a three-phase primary supply line.

With this finding it became clear to me that during the storm when one of the phase wires was broken and came down on A

Street, the feeder fuse on that phase at the D Boulevard tap-off opened up, thus deenergizing it. However, since the remaining two phase conductors were still intact, their fuses had not blown and they continued to be fully energized. That was the reason some of the residences along A Street were entirely out of power while others had severe brownouts, depending if they were fed from the downed phase or from one or the other of the intact phases. The brownouts were due to the fact that the neutral wire had also broken at the same place as the phase wire.

But at the junior high school, which had its three-phase transformers still partially energized by two "hot" phases, there existed a direct circuit within the transformer windings to the third-phase wire that had been broken. Thus this third-phase conductor whose source end was dead, due to its blown fuse, had its other end energized by back-feed from the school's transformers. When I checked on the tap-off fuses to the A Street line, I indeed found that only one fuse had blown and that was on the phase wire that had broken and fallen.

After a study of the many depositions taken from various utility personnel who were active during the storm, we learned how such a dangerous back-feed situation was allowed to remain. The various line crews called in to cope with the storm damage were assigned specific sections of the city in the most affected areas. Also, some of these crews were borrowed from other city electrical systems, near and far, to supplement the local force. The crew that was dispatched to cover the main primary feed line on D Boulevard happened to be a "foreign" crew. However, they were presumably experienced linemen and they had a detailed map of the distribution system for that area with them.

They started their patrol of the feeder in the early evening of the first day of the storm, when the winds were gusting up to seventy-five miles per hour. At the substation from which the D Boulevard feeder originated, they found the feeder circuit must have sustained a massive fault to open these large breakers. They started driving down D Boulevard looking for major damage. Well down D Boulevard they found the massive fault at a large shopping mall. Here the H-framed wood poles supporting the mall's big transformers had buckled, tearing the three feeder line conductors away from the transformer terminals and causing them to fall to the ground—which of course shorted the whole feeder. Since the mall was without power, the crew spent several hours resupporting the conductors and extending them to the transformer terminals again so that they could be reenergized safely.

Then they started back up D Boulevard with the object to return to the substation and close the feeder breakers to restore power to that line. On the way back to the substation, they monitored every tap-off from this feeder going down the intersecting side streets. At each such intersection the tap line coming from the feeder pole was protected by fused "cutouts," one cutout for a single phase tap-off and three for a three-phase line. These cutouts are a type of switch carrying a fuse. When a fuse blows open due to a fault on a conductor somewhere down the tap line, the closing handle on the cutout automatically opens and hangs down. So by observing the cutout handles on the corner poles the crew could readily see if none or one or more of the side-street lines were in trouble or possibly down on the ground. Any tap-off line showing a blown fuse was earmarked for the dispatcher to send a repair crew to pick up the line and restore service.

When the patrol crew reached the intersection at A Street they observed that one cutout of the three there had opened with a blown fuse. To understand their reaction to this finding, note that of the eight or nine street intersections already checked, only one tap-off was a three-phase line, the others being single-phase. And of these the only street showing a blown fuse was on a single-phase tap.

Confronted with a three-phase tap-off at A Street with one conductor of the three probably broken and down to ground, the crew reasoned that as long as the faulty conductor was deenergized due to its blown fuse, it posed no hazard and there was no need to do anything different but to notify the dispatcher. And that was a fatal error they made.

When they arrived back at the substation, they closed in the D Boulevard feeder breakers, thus energizing that whole line, and at the same time by back-feeding as I described previously, they energized with high voltage the load end of the broken conductor lying on the ground on A Street. That was Saturday night and on Sunday morning the fatal accident occurred.

The attorneys representing the contractor's family had no idea of the role played by back-feeding on a three-phase line, but they knew that a supposedly dead conductor could not suddenly by magic, if you will, become alive when touched. They needed a rational technical explanation, and I was asked to look into the matter.

The case, contractor's family versus the public utility, was set for trial in a state court, and in anticipation I was deposed several times by the utility lawyers. At first they would not concede that there was such a thing as back-feed. And even if there was, the victim was at fault for attempting to handle any power line lying on the ground. However, the utility's engineers were obviously aware of a back-feed possibility, but were not saying too much about the training of linemen or electricians to recognize this type of situation.

I have always enjoyed making working models of technical problems and found these to be most helpful in getting nontechnical people to understand what is going on. After I made a working model of the A Street tap lines, the street residences, and the junior high school's three-phase transformer bank and showed without any doubt how the back-feed from these transformers energized the downed line that otherwise would have been dead, there was no further argument about what happened.

The case never came to trial. It was settled by negotiation between the respective attorneys, and the settlement was substantial. Some new lessons were learned by those who should have learned them before.

The Case of the Burning Tunnel

A number of summers ago, my wife and I were attending a chamber music workshop in Northern California. This was an annual event for us as charter members (playing flute and cello). The workshop is a concentrated and consecrated week of music playing in small groups using music spanning the Renaissance, Baroque, Classical, Romantic, and Modern composition periods. Each day a different specific piece, one of many, is rehearsed and then presented in the evening before an audience of some one hundred fellow participants.

A particular gratifying aspect of these workshops is that we are able to isolate ourselves from the world and its disturbing affairs. We read no newspapers, listen to no radio, and don't look at any TV. We leave the local telephone number with a family member only to be used in an emergency.

And so it was more than a little worrisome during a midweek rehearsal session to see our director come in and tell me I was wanted on the telephone in his office—and that he thought the call was urgent. It was!

The call was from an attorney in Seattle. He said he had a hard time finding out where I was. He said that there had occurred a major power failure at a big hydroelectric plant and that several of the large transmission cables were involved. He said he was representing the cables' makers, that they were with him at the dam now conferring with the federal dam operators, and that he wanted me to come up there just as quick as I could. I interrupted before he could get much farther and told him as earnestly as I could that I was some eight hundred miles from the dam site and that we, my wife and I, had driven down to this place and that in addition, and most important, both of us had to play in a musical performance that evening come "hell or high water."

Well, the attorney didn't seem to appreciate the musical commitment, but he did appreciate the driving time difficulty. When calm was restored, I promised to leave the next morning, drop my wife off at home, and then continue up to the dam—hopefully arriving the evening of the second day. The trip did take a little longer than I expected, and it was almost midnight when I got there. Nevertheless, I went right into a meeting with the Japanese cable maker's engineers and their attorneys and I learned the full extent and grim seriousness of the accident.

So that the reader can better appreciate the situation, here are a few facts about this hydroelectric plant. It is one of a number on a major river of the Northwest but is by far the largest. In fact, it was and is now one of two or three of the largest hydroplants in the world.

Four or five years before this accident, in order to increase its power capability while at the same time take advantage of unused excess river flow and head, an additional power house was built of such size as to provide a capability greater than that of both the original existing plants combined. Also, it was decided to transmit this power at the then highest alternating voltage in the country, namely 530,000 volts. This called for special giant transformers to step up the comparatively low generation voltage, special switch gear, special surge arresters, special cables, special overhead line facilities and towers and et cetera. In short, a whole new 530,000-volt operating complex was designed and built by various manufacturers. There were six generating units in the plant divided into two groups, each group having its own transformers and outgoing cables.

The six step-up transformer banks (three in each bank) are mounted on a deck just outside the dam's generating floor. To get the high voltage power out to the 530,000-volt switchyard, nearly two miles away on the other side of the river, it was necessary to provide a transmission means across the river. The plant designers elected to do this by cables passing through and up the various galleries and tunnels that thread within the heart of the dam wall itself. The selected route required about one and a quarter miles of cable length and a rise a little over 650 feet to bring it from the transformer deck at the base of the dam up to ground level across the river.

Within the dam tunnels the cables were hung against the concrete walls on special hangers that permitted expansion and contraction movement as load came on and went off. To take care of the enormous power to be delivered through these cables, each transformer bank fed into a three-phase trefoil cable arrangement, making nine cables for each generation group. Also, to avoid a failure and fire possibility of No. 1 generating group cables affecting No. 2 group cables, the two sets were run walled off from each other all the way through the tunnels.

About the cables themselves, they were a high-pressure oil-filled type insulated with paper tapes impregnated with special insulating oil. The single conductor in each was hollow to permit oil flow into and out of a pressurizing reservoir mounted at the terminal end. In this way the internal insulation of the cable was under a constant fluid pressure, regardless of heating or cooling due to load changes. Of course for the high voltage the insulation was very thick and to withstand the high base fluid pressure a thick aluminum sheath was used instead of a conventional lead sheath. What with all of this the cables were more than eight inches in diameter and very heavy.

So much for background of what this hydroplant consisted. Now about the accident: The cables were energized and supplying power into the northwest grid of the state for a total of only some 350 hours when, about ten days before I arrived there, one of the cables failed in the tunnel at a location near the top of the dam. The arcing at the failure immediately involved the rest of the cables at that location, destroying their insulation and bringing on a holocaust of arcing and fire, which raced up and down the tunnel for several hundred feet. And here is the unbelievable thing! This holocaust of arcing supplied by the enormous energy capability of the hydroplant went on for perhaps ten minutes or more, because there were no quick-acting generator circuit breakers in the plant. None had been provided. The only way to shut down the flow of power was the relatively slow method of steadily reducing the excitation voltage of the generators.

I went into the tunnel and saw the aftermath of the fire. I have never seen such fire destruction to so-called unburnable materials. Every bit of metal in the near vicinity of the failure was in the form of molten pools on a concrete floor that itself looked like rough slag. The concrete walls of the tunnel had been vitrified for a hundred feet into a glassy structure. I was told that the fire fighting crews that responded could not get into the tunnel for forty-eight hours after the power had been shut down. It was indeed a ghastly sight.

I spent the next few days with our chief attorney (who turned out to be very effective) and the Japanese engineers. We went over every aspect of the cable design, manufacture, and proof testing. Also, these same engineers had supervised personally every element of the cable installation. We made many lists of what could possibly have gone wrong, giving priority to two such listings, i.e., possible initiating events, that were the manufacturer's responsibility and those that were the responsibility of the power plant's operation. With these in mind and with concurrence of the power plant's operators and other involved federal agency's personnel we agreed to: (1) take samples of the cables as close to the failure as possible and make detailed dissections and failure analyses based on these; and (2) make a detailed study of the hydroplant's equipment features and its operating procedures before and at the time of the fire.

I should say that at this time the power plant people were convinced that the failure was due to a defect in the cable. This conclusion was arrived at through observation of a small cut-out length that showed that the stiff, rigid aluminum sheath had lost contact with the underlying metal shielding tape over the cable core.

Let me set the scene when we started to dissect the cable samples. We were in a small concrete-lined chamber inside the wall of the dam. There were no windows or other apertures, and the lighting and ventilation had to be supplied through use of extension cords and portable air ducts. There were more than a dozen people crowded into this room: five or six Japanese engineers, each with camera and small tape recorder, our two attorneys, two federal agency representatives, the power plant's managing engineer, two or three electricians from the local work force to do the actual cutting into the cable samples, and finally the two experts who were to conduct the dissections and analyses (a federal agency consultant and I, the cable maker's consultant).

Well, we labored in that crowded, uncomfortable concrete cell for two full days stripping off every shielding component on each sample and layer by layer of the almost three hundred impregnated paper tape layers, studying every possible lay-up irregularity, creasing, contaminant or other blemish by eye and under magnification, photographing what seemed unusual or important, checking for moisture presence, and setting aside fluid samples for later laboratory analysis. My fellow dissector, the other consultant, was looking for defects, and I must say he had an overworked imagination. He was so busy photographing imagined spots that he thought were failure-making incipients that he missed some really important telltale signs of what might have happened. My own conclusion after these dissections was that the cable samples showed no defects of material or workmanship that could have led to the failure. However, with a seven-power glass I did see a scattering of tiny carbonized fibers in butt spaces between paper tape turns. These were located in about a fifth of the insulation wall closest to the conductor. To me, from a background of much experience, this kind of scattered carbonized incipient points was indicative of transient overvoltages on the cable.

In any high voltage transmission system there is always the possibility of surges of extremely high voltage of extremely short duration. Such transients can be due either to lightning or to switching operations on the system. In particular, when a cable is connected to an overhead line the transient generated on the exposed line can enter the cable and pass through its length, often amplifying itself by repeated reflections from the cable terminals. Transient voltages can be in the millions, but they exist at any one place in the cable for only millionths of a second. Depending on the cable, the effectiveness of the surge arresters, and the number and magnitude of the transients, a cable can be totally unaffected by such surges, sustain a scattering of tiny incipient carbonizations, or be so weakened at a critical place as to make a later failure in service inevitable.

So I started to look into the transient situation. Each of the cables as it emerged from the dam wall on the other side of the river was terminated in a big porcelain terminal, and from there the power was carried to the 530 KV switchyard by bare overhead lines on high towers. At the emergent terminal point each cable was committed to a surge arrester, designed to pass the service voltages but divert to ground any significantly higher voltages. A counter at the base of each surge arrester accumulated the number of transients that had been diverted.

On inquiry I found that no records of surge arrester actions were being kept. When I checked on the counters myself, I found all sorts of numbers, from zero to in the teens. But I was told these counters were unreliable.

In the switchyard I found the huge 530 KV circuit breakers at the start of the outgoing overhead transmission lines. These were rated to safely and very quickly open and isolate the hydroplant from the utility system on damaging overloads.

These circuit breakers had been made in Italy and were of very advanced design but had no real record of service. They operated in a sulphur hexafluoride gas and had to be absolutely sealed to perform properly and to prevent any escape of the toxic gas. At the time of the cable failure and fire, one of these circuit breakers, on the same phase as the cable that failed, literally blew apart. There was some evidence also that that circuit breaker blowout immediately preceded the cable failure.

A well-known phenomenon of circuit breakers that are used to protect a high capacitance circuit such as a long cable length is that when they operate open they can at times initiate a series of ever increasing high voltage impulses, which travel into the cable. This then led me to believe that these circuit breakers when they had been operated to open the 530 KV transmission line could allow a rapid build-up of very high voltage pulses. The cables in the tunnel could have been subjected to a number of such overvoltages during their operating period, but the instant when the voltage build-up became so great as to destroy the circuit breaker, it also caused a breakdown in the insulation of the cable and it failed at that time. It seemed to me that this was a plausible reconstruction of the accident because:

1. This was a new circuit breaker design operating at an unprecedented voltage level and having had no previous service experience.
2. My dissection showed no inherent defects in the cable.
3. I found a scattering of tiny carbonized incipients in the inner wall of the cable insulation.
4. I found some indication that the surge arresters may not have been performing their function properly.

I communicated my findings and opinion to our attorney, and it was agreed that these should be presented jointly to the principals in this action as soon as possible.

Then—a blank. For the next three or four weeks I had no word of any kind, either from our people or the hydroplant people. Finally, since I had given up trying to understand what was going on, I called our attorney and asked him "what goes." For the first time he appeared to be evasive and indicated I was to hold everything about the matter in abeyance and he would get in touch with me as soon as he had some answers.

It was quite a while later that I finally gleaned some bits and pieces of information that appeared to indicate that the executive branches of both Japan and the United States had gotten into the act. The result was to call us off all further investigations.

Our law firm was dismissed as well as the consultants, including me, who had worked on the case for both sides. And that was that.

I do know, however, that new cables were installed in the tunnel as well as some new equipment in the power house and switchyard and the hydroplant third-power house is now operating at some level, if not at full rating.

Of course I did feel and even do now feel a good bit of frustration at the outcome of this investigation.

The Case of the Poor Painter

Just speaking about Louisiana calls to mind New Orleans with its exotic Mississippi River setting, Mardi Gras festival, Dixieland bands, and Creole and Cajun cooking. It is also a country of bayous, swamplands, heavy verdure, gorgeous birds, and magnificent trees. But a scant one hundred miles west of New Orleans is a different world. It is a world of oil wells, oil refineries, oil well equipment manufacturers, and oil exploration companies. Here in these oil cities the wealth of "black gold" is obvious—in ornate, massive buildings and many residential mansions.

A few years ago I was called by a firm of attorneys in one of the larger Louisiana oil cities to help them investigate a fatal accident at a refinery there. It was a long trip from my home base and to a part of the country I had never been before. Although the town didn't look like much of a metropolis, it was easy to see that it was humming with vitality. The law offices I was directed to were modernistic and quite luxurious. The attorney in charge of the case I was to investigate was a tall, imposing gentleman whose speech was so Deep South I found it hard to understand him at times.

As he went over the situation, I learned that a painter had been fatally burned in an electrical accident in the power control yard of a large local refinery. It seems that the painter, an employee of the company, was painting transformer tanks in the yard as a part of their maintenance program. One of the transformers was mounted close to a large power control center cubicle consisting of switch gear, metering equipment, and oil immersed overload circuit breakers. Without any warning, the control center exploded, splitting the steel cubicle apart in several places and spraying the surrounding area with burning oil. The painter, not ten feet away, was sprayed with the flaming liquid and was burned over most of his body.

The painter was a young man under thirty and had a family, a wife and several small children. After the shock of the accident had abated somewhat, the wife was pursuaded to seek legal redress. So she engaged a prestigious firm of attorneys to enter suit. She could not sue the refinery, because her husband had been employed there and the state's workman compensation laws limited the employers' liability to medical treatment expenses only. Therefore, suit was entered against the control center's manufacturer, on the basis that its design and construction were inherently unsafe.

The case had come to trial in a state superior court some four months prior to my engagement. At that time the jury's verdict was in favor of the control center's maker. However, the plaintiff's trial attorney had observed during the trial a conversation taking place between the defendant's attorney and one of the jurors. The judge was informed of this and pressed very hard to vacate the verdict and call for a retrial. This he reluctantly did and that was why I was called into the case at this time.

I spent the next two days reviewing the voluminous file of depositions, affidavits, court proceedings, and exhibits. My wife, who was with me on this trip, had an interesting time casing the town and its inhabitants. Her observations, which we discussed each evening, gave us a pretty good idea of how these oil-rich people lived and what they thought about. After I became reasonably informed on the documentary background, I inspected the remains of the burned out control center unit and the actual scene of the accident.

It was an ancient piece of equipment, as such things go in the technical world, manufactured nineteen years ago and in service at various locations for the past seventeen years. Power from the transformer bank entered the control center at forty-eight hundred volts, going to a master, fused blade switch and from there through an overload circuit breaker to the various motor operating circuits. Also contained were current and voltage metering equipment with read-out dials and recording strip charts.

The base of the master switch, made of a hardened plastic material, showed old carbonized discharge tracks between all of the terminal studs. Also, these studs appeared to have been factory mounted unusually close to one another—too close for proper clearance in a forty-eight-hundred-volt system. Another finding of importance was the absence of two of the four bolts that normally held the inspection cover plate at the top of the circuit breaker oil tank. It was obvious that these bolts had been missing for some time—the top edges of the bolt holes showed red rusting.

Fortunately, I had several standard references with me in which the required clearances between energized electrical parts were given. I found that the clearance between studs of the old control center was fully 30 percent less than required for safe operation. Probably nineteen years before, when the control center was manufactured, either there were no national standards or its design was based on what the company thought was safe practice at that time. In either case, it seemed clear to me that a longtime discharge had been going on in that switch between the closely spaced studs. This would have been accentuated during damp weather due to moisture condensation. Then at the time of the accident, the discharge reached spark-over conditions and this was followed by a massive power arcing, which blew the cubicle apart. At the same time, the oil in the circuit breaker tank was ejected violently through the ripped-up inspection plate, caught fire, and was shot out in a widespread flaming spray.

At this stage of the investigation I felt I had hold of some very tangible evidence of how the accident had occurred and who were the responsible parties.

The facts that the switch base showed deep discharge tracking paths that had gone on for a long time and must have been seen on inspection and two missing cover bolts on the circuit breaker oil tank indicates this equipment had little maintenance during its many years of service. An organized program of maintenance both for production efficiency, long life of the equipment, and safety is one of the most important responsibilities of management—and particularly so in an oil refinery, where operational hazards are numerous and high.

The unsafe clearances between energized switch parts was another matter, and that was the responsibility of the control center manufacturer. Even though the standards of nineteen years ago, when the switch was made, may have permitted the close clearances, subsequent disastrous experience by switch-gear users gradually forced them to make substantial changes in their standards, which caused industry changes and wider clearances to be adopted and standardized. Some manufacturers actually recalled the older switches used in forty-eight-hundred-volt circuits and replaced them with wider spaced units in accordance with the later clearance standards. Some manufacturers did not and that was evidently what had happened here.

I expressed my views to our attorney and went over them in detail. He agreed that my testimony in court along the above lines might be a critical factor in the jury's deliberations. So we carefully planned when and how my presentation was to be made.

The trial lasted only three days. At the outset our attorney gave a very dramatic word picture of the young painter at work in the power control yard of the refinery—taking all the necessary precautions, as he had been instructed to do. He had been doing this same kind of equipment painting for the past two years and knew his job very well. And then suddenly there was an explosion and the man was on fire and was burned fatally before anyone could get to him. The defendant's attorneys had brought down four of the manufacturer's engineering staff and put them on the stand one after another. Generally they disclaimed any responsibility for the explosion and pressed the point that they had sold thousands of the old control centers and over the years had had very few major complaints. On cross-examination they were not sure if any were involved in an explosion. Also, they were vague about whether any of their units may have had new switches installed as replacements.

I was on the witness stand for three hours during the second day of the trial. Under our attorney's questioning I gave a straightforward story of my examination of the burned-out control center, the review of current switch-gear reference standards, and my reconstruction of how the accident took place. On cross-examination they hammered at me on two points: 1. In view of the arcing and explosion destruction in the control center, how did I know that there was longtime discharge tracking between the switch studs and that two bolts on the circuit breaker cover plate had been missing before the explosion? 2. If the switch stud clearances were adequate for forty-eight-hundred-volt service nineteen years ago why aren't they adequate for the same voltage service today?

There was no problem in answering the first point. We brought into the court as exhibits a portion of the insulating base of the blade switch and the twisted and crumpled circuit breaker tank inspection cover. On the underside of the switch base the deep carbonized tracks were clearly evident. This underside was flush against the back of the steel control center cubicle and protected from the arcing. The rust spots on the cover plate's empty bolt holds settled the second point.

The most telling point in our argument against the manufacturer was, of course, the unsafe clearance between the energized switch parts. I had with me the industry standards covering such clearances and was about to produce these and read aloud the requirements for forty-eight-hundred-volt service when the judge stopped me.

Judge: What is that you are opening up?

I: This is the *American Switchgear Standard* for 1975.

Judge: Did you prepare that document from your own knowledge?

I *(wondering what he was getting at)***:** No, Your Honor, this standard was prepared by the switchgear committee of the IEEE.

Judge: If you did not prepare it and are not responsible for its content, it is hearsay evidence. Isn't that right?

I *(stammering)***:** No, Your Honor, this is engineering knowledge and is used by all engineers working in this field.

Judge: It is still hearsay evidence, and I won't permit you to give any testimony from it.

Well, that was a blow right between the eyes. If I could not introduce that standard, which required the wider safe clearances, it came down to my word against the manufacturer's word—with the judge obviously biased. As far as I was concerned, the rest of the trial was lost in a haze. Our attorney was seething, but there wasn't much he could do about it. In his sum-up before the jury he went as far as he dared in expressing his legal opinion on what did and did not constitute hearsay evidence. But the judge had the last word. He practically instructed the jury to find for the defendant. And they did.

So much for justice in Louisiana that day.

The Case of the Unbaked Bread

It's my guess that less than one-hundredth of 1 percent of the homemakers in this country regularly bake their own bread and perhaps 10 percent occasionally bake a loaf. At any rate, practically all of us eat factory baked bread regularly. The largest baking companies bake dozens of bread types in addition to their other numerous baked goods, so that even in a good-sized city if one of these major suppliers stops production the shortage is felt very quickly. That was what happened in a Minnesota city a few years ago when a major, probably its largest, baking plant had a complete power failure.

Early on a Sunday morning a passerby saw a pall of smoke rising from a fenced-in transformer yard back of the baking plant. No employees were in sight, so he hurried along the street four or five blocks away to an open gas station and called the fire department. As he put down the telephone receiver he heard an explosion, and when he looked back up the street the transformer yard appeared to be engulfed in flame.

One of the three very large transformers in the three-phase bank had faulted internally; the short-circuited winding had gotten red hot and set the surrounding oil on fire. An explosive mixture of gases from the burning oil then exploded, splitting the top of the transformer tank and spraying the flaming oil over the surroundings.

The fire department trucks arrived on the scene very quickly but had to await the power company troublemen to deenergize the yard before tackling the flames. They soon had the fire out. When it was all over, the exploded transformer was found to be literally destroyed and the other two damaged to various degrees by the fire. Of course, the baking plant was at a standstill, with thousands of pounds of baked goods under way for Monday's delivery, unfinished and unusable.

A baking plant, especially a big one, is an intensive user of electricity if its heating system has been converted from gas. During the twelve or so years of its operation, this plant had gradually converted all of its heating requirements from manufactured gas, the only kind available and fairly expensive, to electricity. To meet the large power requirements, the power company had, during the past year, installed a bank of large twenty-five-thousand-volt transformers, and it was one of these that had faulted and caused the fire. Transformers of this size are not stocked by a power utility but must be specially ordered from a power transformer manufacturer or supplier, with the possibility of considerable delay in shipping. Facing a very substantial loss in revenue and possible plant equipment damage, the plant manager persuaded the power company to provide a portable unit on site to get at least a part of the plant working.

The portable unit was a mobile diesel-driven generator/ transformer trailer with about one-fifteenth the capacity of the baking plant's normal requirements. Power company electricians managed to get the trailer set up in running order by the following Tuesday afternoon and laid temporary cables, openly on the ground, into the electrical control room of the plant. A small force of plant employees had been called back that morning to prepare dough mixes so that by 4:00 P.M. some of the baking ovens were in operation.

Four hours later, the plant electrician smelled burning insulation and on investigation found that the temporary supply cables from the trailer were so hot as to be ready to burst into flame. At that he quickly opened the main power switch in the control room and again all plant operations were brought to a halt.

With this latest happening the utility gave up on temporary measures and concentrated on replacing the two damaged transformer ceramic bushings on the unfaulted units and going after a complete replacement for the destroyed transformer. They were very fortunate to find such a transformer at a used power equipment dealer not too far away, and this was trucked into the plant yard. By Saturday morning the whole transformer bank was back in place, connected and energized. And by the next day a moderate production of baked goods was under way.

Once manufacturing had returned to normal, the baking company took stock of the financial loss during the blackout period and found it was considerable—well into six figures. There was a clause in its power contract calling for a negotiated settlement of any loss caused by an operating defect or negligence on the part of the power company. Based on this, the baking company presented a loss bill to the utility and demanded a covering settlement. The power company countered, stating that the loss was not due to any negligence on its part or to any defect in its equipment or operations. As to the transformer that failed, that they said was due to a longtime heavy, unbalanced loading imposed by the manner in which the plant's electric system served the baking ovens. An impasse was soon reached in the controversy, and the baking company decided to seek a legal solution. Shortly after, their attorney called on me to assist in unraveling the technical aspects of the situation.

Looking back into the plant's power supply record, I found it was one of continuous increase in demand, with the largest jump about two years ago, when the final conversion from gas to electricity had been achieved. At that time, according to the plant's electrical superintendent, the power company had made some suggestions for balancing their three-phase line loadings and these suggestions had been carried out. I also learned that at the time of the fire the plant's main overload circuit breaker had not operated open. Since its overload current setting had not been altered after the new, larger transformers were installed, the transformer failure could not have been caused by any plant overload or short circuit.

Another element of the transformer failure and subsequent fire that bothered me was the fact that the transformer was exhibiting great distress many minutes before it exploded. In his deposition, the passerby who called the fire department estimated there were close to ten minutes between the time he first saw the smoke rising from the yard and the explosion. Granting this and conjecturing reasonably that the smoke came from internal arcing going on in the transformer's high voltage windings, why didn't the utility's high line fuse protecting this transformer open up? And open up well before the explosive phase that destroyed the unit?

The twenty-five-thousand-volt three-phase primary lines supplying the plant came into the yard overhead and were dead-ended or terminated on a pole there. From these terminations wire jumpers went to the bushings on each of the three transformers. At each line termination on the pole there were two protective devices. These were a fused cutout and a lightning arrester. The function of the arrester is to bypass to ground any high voltage surges caused by lightning or switching transients that could damage the transformer. The cutout is essentially a single pole switch that is operated from the ground by a long, insulated so-called hot-stick. The cutout also contains a fuse whose melt-open rating is normally set to take place when the current drawn by the transformer reaches 150 percent of the maximum value designed for that equipment. Also, a cutout has a door or flag that falls open (down) when the fuse has blown and this is easily seen from the ground.

Pressing further into the details of what happened, I learned that none of the cutout fuses had blown open at the time of the fire and the yard had to be deenergized by opening the distribution line circuit breaker at the utility substation located in another part of that area. The significance of this seemed clear. The cutout fuse designed to protect the transformer from overloading was either defective or grossly overrated. As it turned out—and it took a good deal of effort and time to get at the facts—the fuses in all three of the plant yard cutouts at the time of the fire were slow-melt fuses rated to open at twice the current that they should have been set for. That was why the transformer failure and subsequent arcing was permitted to continue uninterrupted as it had. Why had the power company installed such high rating fuses in its primary supply to the plant? As I delved into this question, it became apparent that their distribution department engineers were as surprised as I was.

The answer, however, came when the linemen who serviced the plant's supply lines were deposed. It seems that during a year or so before the fire they had been called frequently to replace fuses in the cutouts due in each case to a few minutes of plant overloading demands. To avoid these unnecessary calls, as they put it, they simply doubled the fuses' size at the last time a replacement had to be made—and that solved their problem. Of course, while it was the linemen's solution, it left the transformers virtually unprotected.

Adding it all together, it looked to me as though the baking plant was largely jusified in making its claim for loss against the power company, because:

1. When the new twenty-five-thousand-volt transformers had been installed, the plant had carried out the utility's suggested changes to better their phase loadings.
2. The plant's main circuit breaker had not opened at the time of the fire (and it was in good operating condition).
3. The transformer that faulted had been in service at the plant for only two years.
4. Incorrectly rated fuses had been installed in the utility's primary supply to the transformer yard.

On the other hand, the baking plant had caused some brief overloading demands to which the original correctly rated fuses responded. It would not be expected, however, that this could cause any permanent harm to the transformers.

In my deposition, taken by the power company's attorney, I went over my findings and the conclusions to be drawn from them. During the next several weeks the attorneys of both parties met to negotiate a settlement that might be satisfactory to both. This they finally did; both companies also agreed and so the case was settled. As I understand the outcome, the baking company received some 65 percent of what it had asked for. I felt this was a fair solution since the power company had to replace one whole transformer and repair damages to the other two, besides losing approximately one week's power revenue.

The Case of the Foolish Fireman

During the late afternoon of a spring day in 1980, the manager of a very large condominium complex in a major north-central city looking out of her office window saw heavy smoke rising from what appeared to be the condo's extensive parking area. Since the corner of one of the building walls obscured her view, she started quickly for the door. At that same instant all the office corridor lights went off—as well as all electric power in the building.

Associating the smoke plume with the power failure, the manager immediately put in calls for the city fire department and the local power company—and then dashed out to the parking lot. There she saw that the smoke was rising from one of the large (500 KVA) pad-mounted transformer cubicles supplying the condo, this one sitting at a corner of the lot near the fronting street.

This transformer, like the two similar ones located on the condo grounds, connected the incoming 13,800-volt primary distribution line voltage to the 120/240-volt secondary service potential required in the buildings. In such a distribution transformer, the utility's well-insulated high voltage primary cables enter the high voltage (primary) side; the high voltage is stepped-down to low voltage, and this appears at the low voltage or secondary terminals of the transformer. It is to these secondary terminals that the consumer's service supply conductors are connected, and in this case these were in the form of buried cables going into the building's electric control panel. It is to be noted here—and this will be of importance later on—that it is invariably the responsibility of the user to select, purchase, install, and maintain the secondary wiring.

Back to the smoking transformer. The manager and the condo maintenance man arrived together and were at the transformer site only a few minutes when the power company trouble truck came barreling into the lot, followed closely by the city fire chief in his private car and then, in another few minutes, two fire trucks.

By that time the smoke pall had become black and very dense and there were now considerable crackling noises coming from the unit itself.

Surrounding the transformer pad was an enclosure made of heavy slatted boards with a padlocked double swing gate also of slatted boards. As soon as the utility trouble man got out of his truck, he strode over to the transformer and unlocked the enclosure gate, swinging both doors wide open. Since it was obvious that the transformer was still energized, he pulled out a well-worn electric distribution map of the area to look for the substation supplying this particular primary circuit. His intention was to use his truck transmitter to call into the load dispatcher to kill the circuit. However, for some reason he was still in front of the transformer cubicle as he was studying his map and with him were the fire chief and the condo manager. There they were then, lined up not ten feet away from a smoking, on-fire, large transformer with its crackling noises of internal electrical arcing audible half a block away.

And the transformer suddenly exploded—with a tremendous concussive violence—blowing out most of the facing windows of the condo and many of those in the buildings across the street. It also killed the three people who were there in front. They were struck by the ripped-out, flying steel sidings of the cubicle. It was a major tragedy.

Eventually the legal actions stemming from the tragedy involved many claims and subrogated claims. Instituting actions were five plaintiffs against eight defendants, with the claims amounting to some $30 million. Involved in these actions were the condo management, the condo constructors and contractors, the manufacturers of the transformer and the cables, the power utility, and the state, through its electrical inspection office.

Each of the litigants called in the services of various experts to help identify the cause of the transformer blowout and at the same time provide, if possible, technical reasons promoting their particular positions in the case.

All sorts of ideas were put forward by these experts on the cause of the explosion. Most of these centered about either defects in the transformer or seepage of illuminating gas into the cubicle—leaking from several large gas mains buried in the vicinity.

I was called into the case by one of the defendants almost a year and a half after the accident. This defendant believed the transformer wiring had become defective and this initiated a sparking that led to arcing and then the explosion. Since wire and cable insulations were involved in his theory, he wanted a cable expert to assist his investigation. At that time I was conducting a cable failure clinic at a nearby university. He visited my class and after listening to what was going on, he thought I could help him. I agreed and so became involved.

By that time the number of depositions, affidavits, and general correspondence had reached staggering proportions. Also, all remaining parts of the transformer, its enclosing steel cubicle, and the buried secondary cable remainders had been removed and stored in a warehouse. Here all the salvaged parts and pieces were laid out on a long floor and were available for examination by authorized personnel with the court's permission.

I spent more than a month reviewing the documents, examining the warehouse materials, visiting the accident site, and talking with the utility engineers and the city electrical inspector. The utility engineers assured me that leaking illuminating gas was not the cause of the explosion. Within two days of the accident an independent laboratory had taken earth samples around the condo's parking lot and analyzed the effluent from these for gas. The utility had hired them to do this. No illuminating gas was found.

Another factor that was somewhat unusual was the loss of all electric power to the condo building while at the same time the transformer remained energized. The only explanation for this would be that the secondary cables serving the building had been shorted out or burned out but that the primary transformer fuse was rated so high that it could continue to function, passing high short-circuited current levels into the fault.

Laboriously reconstructing the original power supply set-up, I learned that there was more to the cubicle than just a transformer housing. On the secondary side the cubicle contained a large distribution terminal board with sixty separate connection studs, all fed from the transformer's heavy secondary terminals. Attached to these sixty low voltage studs were sixty secondary cables, all bunched together as they dropped out at the bottom of the cubicle through a short pipe and then direct buried in a shallow trench some twenty-five feet before entering the building. This solid mass of secondary cables, forming twenty three-wire 120/240-volt circuits, violated safety code requirements. They should have been run in separate conduits with not more than two or three circuits per conduit.

Well, I had discovered the cause of the explosion, and it was a messy thing indeed:

1. Because a local electrical contractor wanted to save some money in connecting the big transformer to the condominium, he packed the sixty secondary cables from the low voltage side of the transformer all together and ran these as a solid bunch in a shallow trench to the building.
2. A year later, a condo landscaper decided to plant a tree right over where this mass of cables was trenched—and one of the cable's insulation was nicked badly in the process.
3. The power utility fused its transformer on the primary, high voltage side with such a large rating fuse so that thousands of secondary fault current amperes could flow without deenergizing the transformer.
4. The stage was thus set for the explosion.

The damaged cable in the presence of moisture started to spark to earth. This continued and finally led to some arcing, and in time this spread to all of the rest of the packed together cables so that suddenly there was a maelstrom of arcing and all power to the building was disrupted. The heavy arcing also pyrolized the polyethylene insulation of the cables and melted all the conductors together. At this stage the liberated gases from the burned insulation—mostly hydrogen—flowed along the trench and up into the transformer cubicle. There, as the gases met sufficient oxygen, an explosive gas mixture was formed, which was then detonated by the sparking and arcing going on in the overheated connections of the cubicle secondary terminal board.

I remember the rather dramatic conference at which it was arranged that I present my theory of cause. There were a dozen attorneys present, each with one or more assistants and experts, insurance representatives, the city electrical inspectors, and several engineers from the power utility. I had prepared some charts and slides to provide better understanding of the technical aspects, and for one solid hour I held the floor, reconstructing the setting and the event-by-event culminating in the explosion. Then for another three hours I responded to countless questions from every litigant present. When it was all over there appeared to be a consensus, unspoken of course but nevertheless present, that the explosion had to have been brought about as I described it.

The suit against my defendant was dropped. However, as I learned later, four litigated suits eventually reached the courts and these ran on for more than two years. As it turned out, the suits against the city failed—much to my surprise. There is no doubt in my mind that had the electrical inspector's office fulfilled its proper commitments and ordered the wiring contractor to install the secondary cables properly in separate conduits, there would have been no explosion.

Leaving the question of the explosion, it has always been a great mystery to me why the three victims of the accident had placed themselves in such a hazardous position. Two of these people were trained professionals, one an experienced electrical serviceman, the other an experienced, active fire fighter. Both of these men must have understood the importance of keeping themselves and everyone else well away from a smoking, arcing transformer until the power had been shut off positively and the unit cooled down.

There is no answer to these fatalities.

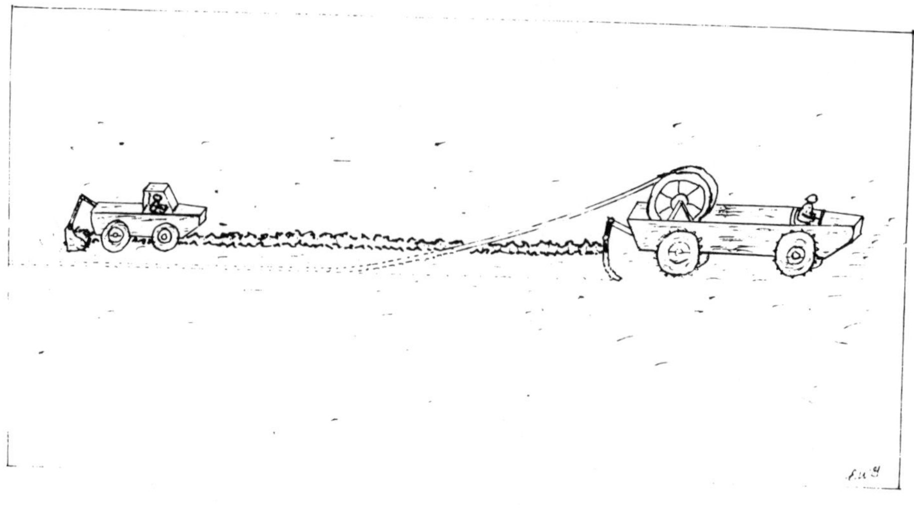

The Case of the Rough Plowman

There are many large alfalfa, grain, and legume farms congregated in the Columbia Basin. Here the proximity of the Columbia River and the usually high water table makes effective irrigation possible—resulting in quickly maturing good crops. This is in contrast with the more northerly plains, where dry farming of grain crops must be practiced.

In the Columbia Basin, pumping stations to supply water to the many irrigation systems in use are scattered all over. Providing electric power to these pumps requires relatively long lines from local utility substations, and to do this economically the lines must be at much higher voltage than required by the pump motors. The usual practice is to carry three-phase primary lines at fifteen thousand or twenty-five thousand volts out to the pumping stations and there transformer banks step the voltage down to the 240 or 480 volts for motor operation.

Formerly these high voltage distribution primaries consisted of bare wires strung on insulators on the cross-arms of poles. Many pole lines of this kind traversing the farm areas always caused interference with the layout of the crops and with farm machines moving around the crops. So when an economical underground cable and installation procedure became available, the utilities changed their distribution system from overhead pole lines to underground cables.

What made this possible was the development in the early 1960s of the UD (Underground Distribution) cable. This was a simplified power cable capable of operating up through the twenty-five-thousand-volt classification using an extruded plastic insulation, chiefly polyethylene. This cable has no metal shielding tapes nor the usual heavy lead sheath but only a helically laid neutral of small copper strands. Its comparatively light weight lends itself to relatively easy, rapid underground installation. In fact, laying UD cable underground in a rural environment became almost as rapid as stringing overhead bare wires on a pole line. The technique developed was to use a bulldozer plow with a large digging tooth to open the ditch to the desired depth—usually thirty-six inches. At the same time, the cable is paid out into the ditch from an axle-mounted reel on a stand at the back of the plow. Following up the plow, a backhoe then backfills the ditch smoothly to grade level.

Plowing-in UD cable in the irrigation fields of the Columbia Basin became the accepted method of getting electric service out to the pumps. Many thousands of feet of cable were undergrounded in this manner without major problems—that is, not until several years after the practice was started.

Then one early spring I was called by a large county utility to investigate the cause of a number of failures they were having with their UD cables in the irrigated fields. These failures started occurring after the cables had been in normal operation for from three to four years. To get the background I needed, I spent a day at the offices of the utility, talking with their manager and several engineers. I learned that they strongly believed the failures they were experiencing were due to defective cable. I came away with the feeling that they hoped I would just confirm that these cables had built-in manufacturing defects and I would not be doing any specific investigation. But *I was* very interested in making a thorough investigation.

To back up a bit, with the development and acceptance of UD cable the nation's power companies started laying millions of feet in new residential developments. Inevitably failures did occur, some in the first year of service and progressively more in each of the succeeding years. Most of these early failures were obviously due to mechanical damage from bedding rocks or inadvertant dig-ins, but a substantial number of the failures were true insulation blowouts—seemingly caused by intrusion into the polyethylene of minute amounts of moisture. This was cause for considerable concern among cable engineers, and as a result, there was much ado about carefully investigating every UD failure. That was why I decided to make a thorough check into these irrigation cable failures.*

Looking over the installation and operating records of the cables that had failed, I was immediately struck by the fact that every one of them had been laid in the ground either on a January or February day. A check of the daytime weather records for these months over the past five years disclosed temperatures had generally ranged from twenty-eight to forty-five degrees Fahrenheit. In answer to my question why cables were laid in such cold weather, the utility answered that the irrigation usually got under way in early May and it was necessary to start putting the electrical distribution system together several months in advance.

I decided to have a close look at a number of the cable failures—both at the faulted regions and also along adjacent lengths. Amongst the ten or twelve failed lengths I dissected were three lengths that showed very little postfailure burning, and I was reasonably able to see the nature and progression of the fault through the insulation to the conductor. What I found was a type of failure I had never before seen in polyethylene insulation. And certainly these were definitely not water-treeing failures nor were they due to dig-in or rock damage.

As you may know, polyethylene is fairly translucent and you can almost see through a considerable thickness of it. Over a two- or three-inch length, the insulation appeared to have a long crack running from the outside all the way down to the conductor, with

*Much later, after six to eight years of continuous service, failures in polyethelene insulated UD cables all over the world reached epidemic proportions. These failures were all of the water intrusion type and came to be known as "water-treeing" failures. A great deal of study and research has been and is being expended to mitigate this disastrous effect.

other smaller cracks at right angles to the main crack. These cracks were readily apparent in all three of the cable lengths. Within each of the cracks were fine, short carbonized tracking paths leading from the outside, from crack to crack, and so on to the conductor. These carbonized discharge paths obviously would be an initiating phase for a final, massive strike-through puncture. Fortunately, the final failure stage had not occurred in these samples.

Under a twenty-power microscope, the cracks appeared to be cleavage planes along crystal boundaries in the insulation. Now a hot extruded material like polyethylene insulation would seem to be completely free of crystalline structure, but actually there is always a substantial crystalline content in all extrudable plastics and polyethylene is no exception. The question in my mind regarding these cracking failures was, What could have caused a bond weakening and sliding of the polyethylene material along its crystal interfaces at those particular places? Then I recalled that crystalline fractures in all materials occur most frequently when the material is stressed at low temperatures. Following this, the fact that these failures all occurred on cables that had been laid at low temperatures assumed great significance. It seemed almost certain now that something was happening during the plowing-in operation in January and February that caused fracturing of the cable insulation.

Checking into this, I learned that the utility hired a local heavy-earth-moving contractor to do all of its cable laying in the irrigated fields. Also, this contractor was given verbal and written instructions on preferred laying procedures, and in addition, a utility inspector was to be present during all laying operations. He was—but only during the first year. After that, the utility reasoned that the contractor's experience justified eliminating the need for an inspector. As it turned out, this was not a good decision.

When it became apparent to the utility that my investigation was taking the responsibility for the cable failures further and further away from the manufacturer, they became less cooperative in providing information and opportunity for detailed observations. However, I finally convinced the manager that regardless of where the responsibility lay, it was of first importance for the utility to find out why these cables were failing on their system.

Well, the answer came rather quickly. A day spent talking

with the contractor's plowing crew showed clearly how a new kind of rough treatment in the laying operation could cause a new kind of cable failure. Briefly this is what happened: the pay-out reel—that is, the reel of cable mounted at the back of the plow paying out the cable into the ditch as it was being opened up—*had no braking action.*

Normally a so-called brakeman, in attendance as the reel is rotating, controls the rate at which the cable is being paid out to keep it in step with the forward movement of the plow. This is necessary to provide a smooth, straight lay-out of the cable without any undue tension. Also of great importance, the brakeman can immediately stop the pay-out reel when the plow stops moving. If there is no braking action, when the plow, which has been moving forward at a good pace, suddenly comes to a stop, the cable reel, which has been rotating rapidly, will continue to pay out a large slack footage of cable that will wander anywhere and everywhere over the ground. Then as the plow begins moving again, it starts taking up the haphazardly laid out slack—at first slowly, then faster. As the last of the slack comes out, the cable suddenly receives a strong whiplash or snap, the degree of which depends on how heavy the pay-out reel is at that time.

It was this whiplash of the cable during the cold weather that produced the observed cracking in the polyethylene insulation. There was no doubt of it, as the same phenomenon was duplicated under laboratory conditions.

So I did not help my utility client to demand replacement of defective cables from the manufacturer. But they did learn how a crystal cleavage failure definitely could be avoided, and that was even more important.

Since I believed I was the first cable engineer to discover this type of cracking failure, I wrote a technical paper describing the fault and how it could be investigated. The paper was published under the title "Fracture Failure in Polyethylene Insulated Cables."

The Case of the Big Intake

San Francisco, that wonderful motley of people, buildings, places, and things, aptly characterized as Baghdad by the Bay, is the most cosmopolitan city in the U.S. It is a city of contradictions, with its mysterious Chinatown and its little old cable cars running up and down in the heart of downtown. A few blocks away are the tallest, most massive office buildings in the West. And a few more blocks away is the new land reclaimed from the bay, already covered with buildings and commerce.

Down in that area, near the shipping wharves, a towering new skyscraper was just nearing its break-even space rentals when suddenly and with no prior warning ten floors of office space became untenable and occupants were forced to vacate.

It was during the afternoon of a busy weekday that a fire broke out in a transformer vault situated twenty feet below the pavement in front of the skyscraper. It was not a fire in the usual sense, since there was very little structure to burn in that vault. However, some electrical equipment had faulted and there was considerable internal arcing going on, which melted open the steel tanks of the transformers and other units. These were filled with insulating/cooling liquid, which under the four-thousand-degree-Celsius arc temperature evolved vast quantities of smoke. This pale, yellowish smoke poured up out into the street through the grilled vents in the pavement and continued to do so after the fire alarm brought the city fire trucks and the power company troublemen to deenergize the vault.

Loss of electric power to the area was over quickly, since as in most high density commercial sections, it was supplied by a secondary network. Such secondary networks furnish power to the buildings at their utilization 277/480 voltages but are supplied from more than one primary circuit. Therefore, if a fault occurs on one of the primaries it is automatically disconnected from the secondary network and the other primaries continue to supply the network. So loss of power was not a problem. But the smoke from the vault fire *was* a very serious problem because it contained PCBs (polychlorinated biphenyls).

High voltage transformers, circuit breakers, and similar equipment operate in sealed tanks filled with a suitable liquid. Conventionally this liquid has been a petroleum product, a thin highly insulating and cooling oil. However, oil can burn and carry flame. During World War II an insulating liquid was developed for the armed services that would not burn, and this came into wide use by the federal government. After the war, the benefits of a noninflammable transformer and switch-gear liquid were publicized industrywide. The new insulating liquid was given the generic name askarel. It was listed by the American Standards Association, and its use spread rapidly in both private and public agencies.

While an askarel does not flame, it does decompose at high temperatures, liberating a heavy, vaporous smoke. Depending upon the grade of askarel involved, the smoke consists mostly of hydrogen chloride and PCBs. During the past decade it was determined that PCBs are toxic and cancer-producing, and the national Environmental Protection Agency has banned the use of PCB in new equipment and issued stringent regulations for the handling and disposing of presently installed apparatus containing an askarel.

The network equipment in the vault fire contained hundreds of gallons of askarel liquid, and the smoke that poured up to the street was rich in PCBs. At the street level, only a little offset from where the smoke was rising, the skyscraper had a large decoratively grilled air intake for its air-conditioning system. This grill, at least eight feet high by six feet wide, was a work of art in bronze. During the entire time that smoke continued to come up from the vault, it was being sucked into the building's air-conditioning ducts. Of course the intake system was protected by filters and a smoke sensing alarm. However, the filters removed only particulate matter and the smoke sensor for some reason did not operate. That is how this comparatively new building with offices and corridors all equipped with air-conditioning vents supplied by the street level air intake became contaminated with PCBs.

As soon as the presence of PCB contamination in the building was verified by chemical tests, the offices were vacated and closed floor by floor, up to and including the tenth.

The building owners demanded that the power company immediately engage expert facilities to decontaminate every space in which PCBs had been found. The power company agreed to do this, but after spending many millions of dollars with only moderate success, they gave up the job. Most movable objects on these floors had been taken out and decontaminated or destroyed and the bare walls, ceiling, and floors worked on—but it wasn't enough to yield negative chemical tests.

As matters stood when I was called in several months later as an expert to determine the cause of the vault fire, the vacated portion of the building was still empty. The principal leasee was suing the power company and the building owners, while the building owners were suing each firm that had any participation in the design, construction, testing, and inspection of the building. In all there were fifteen litigants active in the various suits and countersuits. In view of the legal complexity, there was talk of appointing an arbitrator to hear and make binding decisions concerning each litigant. I don't know if this was done.

All of the power company's network equipment in the vault fire had been removed and impounded in a segregated warehouse. I received permission to view what was there and spent most of the morning doing that. Each of the large network pieces was completely enclosed in heavy clear plastic, and the cables and other smaller parts were under glass in cases. While these precautions against contact were commendable, the coverings on the equipment I came to see did prevent good seeing. I found three sizable network transformers but only two network protectors, along with a great quantity of badly burned cables. Of the network units, two of the transformers looked to be in good condition. The third transformer was heavily damaged by arcing, with most of the burnout on the secondary side.

Although I had no background information from the utility on the operation of this vault's network supply (and still don't have any), it seemed to me that the fire was due to a malfunction of the network protector in this particular transformer circuit. Since in a secondary network system there are many primary feeders, if one primary gets into trouble and is deenergized there is a good possibility that the secondary network will feed back reverse current into the deenergized transformer. To prevent this a network protector is inserted between the transformer and the secondary network whose function is to open the transformer's secondary when it detects reverse current flowing from the network.

Such a situation may have occurred in this case. The transformer primary was deenergized for some reason—perhaps a primary cable fault or a fault in the transformer itself. When that happened, reverse current from the secondary network started to flow back toward the secondary side of the transformer. But the network protector failed to open the circuit to prevent it.

The consequence was that the full power of the secondary network invaded the deenergized transformer, caused its secondary to short-circuit, and reenergized the primary winding in a step-up rather than a step-down action and the arcing that followed destroyed the transformer. The greatest destruction would be on the transformer's secondary side—as actually found.

I discussed the above ideas about the fire origin with the attorney for the company I was assisting, stressing they were very preliminary. I also gave him a list of technical questions that required answers from the power company and a request as to the whereabouts of the third network protector.

For reasons unknown to me, my request for information was never answered. I did hear in a roundabout way that the power company was replacing network protectors at other locations with new units. The removed protectors I was told had all been purchased at the same time as those in the vault fire. Also, the rumor was that there were metal shavings left in these protectors during manufacture. If true, the above information bolsters my opinion of a defective network protector as fire cause.

I don't know how the legal situation was resolved or if it was resolved at this writing. As too often happens to a consultant, when the client feels the consultant's usefulness is at an end, he just terminates all contact without further notice. That is what happened in this case.

I do know that the building has now been fully reoccupied. However, it is my guess that the legal problems have not yet been untangled.

The Case of the Lost Sunday Dinners

The area north of Seattle, Washington, has experienced a tremendous growth of population over the past decade. Scenic beauty, mild winters, the proximity of all kinds of water sports, boating and shipping facilities, and, most important, the location of many employing industries, such as aircraft, lumber, paper products, and plastics, are important factors. Also, excellent transportation has been provided into and out of the booming metropolitan sections of Seattle.

Keeping pace with this growth, the electric power facilities have had to double and redouble their service capabilities for the ever increasing industrial, commercial, and residential loads they faced. Most of these utilities purchase their electricity from the federal agencies that operate the great hydroelectric dams in Washington, and that poses no problem. However new, enlarged, and extended transmission lines and hundreds of miles of new and extended distribution facilities have been the real problem—particularly the latter.

Adding to the distribution concerns was a state enactment requiring new residential supplies to be underground. This meant that the usual 15 or 25 KV primaries entering a new development had to be in the form of buried insulated cable, the transformers supplying the 120/240 volts to the residences had to be in an underground vault or on a suitable surface pad, and the service wires into the house had to be in the form of buried secondary cable. This was a much more expensive arrangement than the usual pole-mounted bare primary running to a pole-mounted transformer and then going as a hanging service drop into the house.

To offset some of the primary power cables' high cost, the cable manufacturers together with certain national utility engineering committees developed a simpler, scaled down power cable version. This type, known as U.D., has a polymer insulation (generally polyethylene, PE, or ethylene propylene rubber, EPR) and a light, compatible extruded jacket loaded with carbon particles rather than a metallic shielding tape. The carbon particles in the jacket are there to make the extruded plastic jacket sufficiently conductive to act like a metallic tape shield. Such a conductive shield is required by code for all cables used at voltages over 5 KV.

In the early days of manufacturing UD cables, many types of carbon and its inclusion in the plastic during extrusion of the jacket were tried. Some appeared quite good, mostly lamp blacks, but others were unusable since they provided little electrical conduction. My story of the "lost Sunday dinners" involves a large quantity of installed UD cable in which there was little or no shielding action by the carbon loaded jacket. But let me back up a bit.

The power company of which I write had ordered many thousands of feet of this new UD cable insulated with EPR to provide a service to at least a dozen large residential developments that were under construction. As the developments were completed, the UD cable and the transformers were connected up and service power became available to the new residents. A few months later, as another group of developments neared completion, more UD cable was ordered and installed.

Everything went along fairly well for the next six or seven months, and then—bang! Some of the UD cables began to fail in service. These were dug up and replaced, but an attempt made to determine the cause of failure produced nothing definite. Then the failure rate increased to epidemic proportions. During the next six months the utility experienced 132 failures, of which 40 occurred in one month alone. By that time, the cable manufacturer's technical representatives were literally living in the area, examining each installation and dissecting many of the failures as they occurred. However, they stopped short of declaring their cables defective.

In addition to the loss of revenue and the expense of continually digging up the failed cables and replacing with whole new lengths or splicing in short lengths, the utility's customer relations deteriorated badly.

Oddly enough, many of the failures occurred on a Sunday at an hour or two before noon. Since the residences were all-electric, this meant that many Sunday dinners were lost, and of course all blame was heaped on the utility.

During the first rash of failures, replacements were made from a considerable warehouse stock on hand. But as failures multiplied, the utility purchased emergency supplies of UD cable insulated with polyethylene from another manufacturer and used this for replacements. They were convinced that the failures were due to some inherent defect of materials or manufacture in the cables. On the other hand, the manufacturer's investigators were still looking for system or installation problems as causes of the failures.

Somewhere at about this stage of the troubles, I was called by the utility to find out what was going on with the UD cable. I turned part of their machine shop into a laboratory (over some protests) and set up a few pieces of equipment and instruments to dissect and analyze UD cable failures. And I was deluged with them. In three or four days I had examined some twenty failure samples, and by this time I knew what was wrong with the cables.

I mentioned something about the UD cable design before; here is a bit more detail of its construction. Although a fifteen-thousand-volt cable, it had a relatively thin wall of insulation because the material used, EPR, like polyethylene, has exceedingly high electrical strength. As a protective cover for the insulation a jacket was provided, which was made conductive by loading it with suitable carbon powders. However, at the time the UD cables were made this jacket was of a different ethylene compound than

EPR, because that manufacturer had not yet worked out the technique of incorporating the amount of carbon needed into an EPR resin, but he could in the ethylene compound. Over this jacket were a number of tin-coated copper (neutral) wires spiraled around the cable. These wires took the place of the usual heavy, expensive lead sheath in a regular power cable.

For satisfactory operation of a UD cable the conductive jacket has to be tightly adherent to the underlying insulation and maintain a high degree of conduction. The two common defective features of these cables I found during the dissections were:

1. The jackets were mostly loose over the insulation, and as such, there was evidence of electrical discharges and burning.
2. In every failure sample the jacket was found to be nonconductive.

So the UD cables originally installed by the utility had built-in defects that could shorten their service lives drastically, as they did. Although originally manufactured with the jackets in tight contact with insulation, since they were not of the same materials, there was no bonding and after heating and cooling under service loading they were bound to separate because their thermal expansion characteristics were different. In addition, although as manufactured the carbon in the jackets provided the necessary conductive properties, there came about a gradual reduction of conductance with time. After six months, for example, there was practically none left—meaning that the cables were operating essentially without shields and this subjected them to concentrated stresses, which can cause failures.

All of this was presented to the cable manufacturer with the demand that he provide remuneration for removing all of this type of cable from the housing developments and pay for new cable (not his) and installation of same. Also, he was to take back for credit any of his UD cable remaining in the utility's stocks. The manufacturer agreed partially. He would replace all of the defective cable, but with his own cable, this time using polyethylene insulation and a polyethylene conductive jacket (same materials). Also, he would not cover removal and installation costs.

So the matter went to court for settlement of the issues. As the trial proceeded, it was evident that the nature of the defects and how these led to the cable failures was not in serious contention. The degree and amount of damages was the major issue. Finally the case was handed to the jury after the judge had instructed them as to the specific elements they were to deliberate on in reaching a verdict. The utility's position was sustained, and an amount of damages was awarded—not all that was claimed, but still a reasonable amount.

Something of real interest to me came out of the evidence presented at the trial. I had been speculating about why the originally highly conductive jacket on these cables had lost most of its conductance after only several months. During the trial one of the cable maker's engineers said that their jackets had adhered so tightly (originally only) to the insulation, they were concerned that it could not be removed (as it has to be for a short distance) at the joints and terminals. To make it strip off more easily, they had inserted a so-called breaker compound over the insulation before extruding on the jacket. On our question he indicated that this compound contained a large proportion of acetic acid.

That was the clue to what happened. It is now well known that acetic acid has an affinity for carbon and will react in such a way with carbon particles as to form them into clumps. Of course this breaks up the carbon-to-carbon conduction pathways and after a while the jacket will show no conduction—as we found.

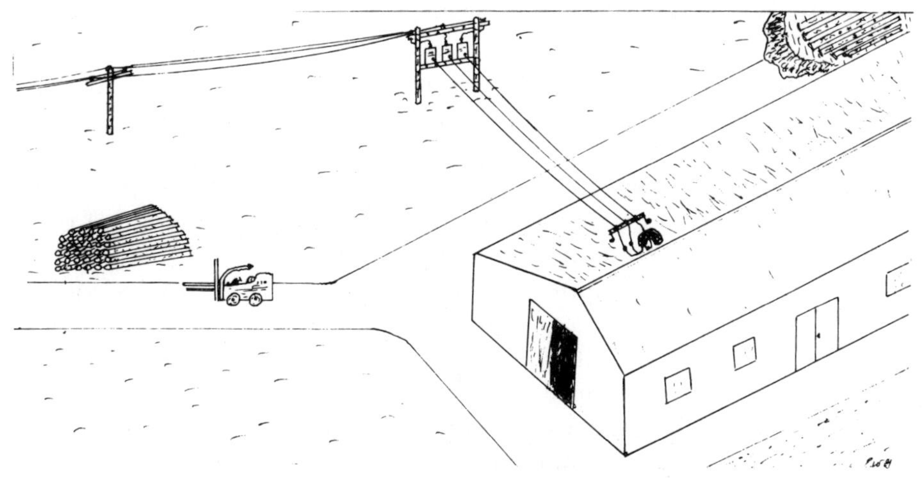

A Case of Mistaken Identity

In some states, particularly in the Far West, a single large electric power utility operates throughout the length and breadth of the state. Where that happens the major executive and engineering offices are located in a principal city and small service shops are scattered around the rest of the state—often quite far-flung. Service calls for power failures and for minor line construction can be handled adequately by these service shops. However, if called upon to provide new construction requiring technical design or technical analysis of system problems, they have to refer these to the engineering offices at headquarters. Sometimes such long-range action between the "know-how" and the "doing" entities can lead to misunderstandings and even serious consequences. This tale is a case in point.

A few years ago, in the Northwest's heavily timbered areas one of the large plywood manufacturing mills found it expedient to undertake a large expansion and complete modernization program. This mill was located close to the state line in a relatively remote area. Eventually the local power service shop was contacted and asked to provide the much greater power required by the added facilities.

In order to do this, the power company was given a detailed list of all the planned motor, heater, lighting, et cetera, loads and an estimate of how the mill's electric demand would vary during the day and during the week. This information was passed on by the local service shop to the power company's engineers. They calculated the new energy requirements, checked on the local substation and line capability, determined the size and type of the new transformers and service drops to the mill, and then wrote out detailed specifications for accomplishing what was needed. This was passed on to the local service shop to carry out the construction and installation phases using the materials furnished from the main warehouse at headquarters.

As the mill's expansion work neared completion, the new power supply was put into place, tested, and turned over to the plant's electricians. Three weeks later the new plywood facility went into operation, and several weeks after that the very modern specialty products unit was also in production—both at full capacities.

Almost three months to the day from start-up, the plywood mill caught fire and it was literally burned out in a short time. Arson was suspected at first, and the state's fire marshal was brought down to investigate the fire. He traced the fire origin to the roof of the plywood mill and to the specific part of it where the mill's electric power service entrance had been located. With this positive identification of where the fire had started the arson theory was dropped and the finger was pointed at the power company.

After a considerable discussion between the utility and the mill owners, the situation boiled down to the following contentions:

1. The mill owners and their insurance company were convinced that some misdesign or defect of materials or construction of the new power supply caused the fire.
2. The utility categorically stated their power supply was in every respect in accordance with national safety codes and disclaimed all responsibility.

As a result, the mill insurers entered suit against the power company to recover their damaged property and business loss. Later their attorneys called on me as an expert to examine what remained of the mill's electric power supply components and, they hoped, to develop the actual details of how the fire got started. And so it was that I donned a battered pair of overalls, a stocking cap, and a filter respirator and spent two days rooting in the charred, scattered debris of what had been the plywood roof. It was hot, dirty work, but it was rewarding. I found that the power company had made two mistakes—both involving their service drop cable. This cable went overhead from a pole platform where the new transformers were located to the weatherhead service entrance pipe on the roof going down into the mill's electrical control center.

Mistake One: The service drop cable supplied by the utility had *aluminum* conductors. On the other hand, the weatherhead service entrance, which was part of the mill's construction and owned by the mill, had cables with *copper* conductors going into the mill. For some reason the local power shop had used bolted copper connectors to join these two different types of conductors together. That inevitably led to the bad situation known as thermal ratcheting, which after repeated cycles of load off and on causes progressive loosening of the aluminum conductors in the connectors. Finally the connection becomes so bad that the heat generated can bring it to red hot condition—so that the aluminum conductor strands can melt. What I found indicated that this indeed had happened.

Mistake Two: There were a number of spots of congealed, blackened plastic blobs on the charred roof members that were originally right under the weatherhead pipe. I dug up some of these blobs, quartered them, and looked at them with a five-power binocular glass. They were undoubtedly blobs of polyethelene—the insulation of the aluminum conductor service drop cable. That made me examine closely the rest of the intact service drop close to the transformer pole. The insulation *was* polyethelene. The next step was to find out if it was the thermoplastic type or the thermosetting type. This is important because the thermoplastic polyethylene melts at a relatively low temperature and burns, carrying flame. The thermosetting or cross-linked polyethelene will not melt or burn; when overheated it just chars and cokes, giving off gases. My quick test was to cut off a piece of the service drop insulation and put a match to it. It caught fire at once, and as it melted, it dropped flaming blobs. Obviously the power company had supplied the mill with a service drop cable insulated with thermoplastic (burnable) polyethelene.

After these findings, reconstruction of the plywood mill fire origin was simple and definite. During its three months of operation, the mill had been producing at top capacity—meaning its electrical loading was near maximum. The thermal ratcheting going on at the three service entrance connecters was progressively loosening and heating the aluminum conductors there. The last phase was a jump from moderate heating to "red-hot" stage, and at this point the polyethelene insulation caught fire and dropped flaming rivulets of plastic onto the roof. It caught fire and soon the wood roof itself started to burn.

In my deposition, taken by the power company attorney, the above findings and deductions were brought out. This led to several informal conferences between the principals and their attorneys. We learned that the power company engineers had specified an aluminum service drop, not knowing the mill had copper service entrance conductors. Also, the local service shop when confronted with making connections between the two different conductor metals chose the wrong kind of connector. Also, the utility engineers had specified cross-linked (thermosetting) polyethelene insulated service drops. However, both types of insulation look alike, and unless the jacket printing is clear and is *read* it is very easy to mistake one for the other in the field, and that is what happened.

On the other hand, we learned that the mill's roof construction was not fireproofed to the degree that it might have been. This, the utility contended, led to the unusual rapid spread of the fire.

The upshot of these meetings was a negotiated settlement more or less satisfactory to both parties. A benefit, if it can be called that, of the fire was the lesson learned the hard way by both the power company and the mill.

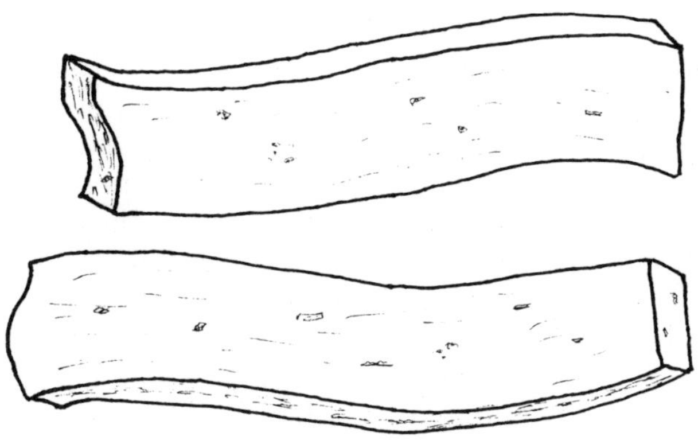

The Case of the Twisted Bars

Our Northwest is a land of great scenic beauty. It is blessed with a rugged ocean coast, majestic mountains, great forests, and many large lakes and rivers. In particular, the mighty Columbia River entering at the Canadian border drops steadily with many twists and turns the breadth of Washington and then turns west in broad stream and moves through deep gorges to the ocean. Its traverse covers almost fifteen hundred miles.

During its run, the Columbia main stream supports some thirteen major hydroelectric plants, one of which up to a few years ago was the world's largest. And since production of electric power by hydroplants is much more economical than by other means, the Northwest has become a mecca for industries requiring very large amounts of power in their operation. Probably the most power-hungry industry is the manufacture of aluminum metal, and most of the nation's aluminum production companies have huge plants there.

Metal production takes place in a large carbon-lined pot in which the oxide ore is reduced to metal by the flow of heavy electric current (d.c.) in the presence of a catalyst, fluorospar. It takes approximately eight thousand watt-hours of electric energy to produce one pound of aluminum. The typical pot-line consists of many pots in series, each taking approximately six volts of the voltage applied to the line. Current flow through the pots can be in the thousands of amperes so that very large, massive busbars are required to electrically connect the pots together and to the power source. Aluminum, second only to copper as a practical conductor of electricity, is used for the pot-line busbars, and these can be typically several inches thick by several feet in width.

Such bars are cast from high conductivity, pure aluminum metal in relatively long lengths. Then fabricating the cast lengths into the configurations required by a given pot-line means cutting to size, deburring, bending, drilling for jointing bolts, et cetera—all to engineered specifications. And thereby hangs my story of the twisted bars.

One of the big aluminum companies of the Northwest was in the midst of a major expansion about ten years ago. They had planned on adding two complete pot-lines to double their production rate, and construction of these was sufficiently along so that the busbar system could be designed. The company then had the necessary metal in the form of busbars cast at their large eastern facility and shipped out to the plant undergoing expansion. At this stage the management decided to seek a local source on contract to fabricate the rough bars into the finished pot-line busbars.

As it happened, the A.J. Company, a small electrical contractor in eastern Washington, won the bid for this work. This was by far the biggest venture ever undertaken by A.J. A.J. had no facilities at the site of the aluminum plant nor much of the heavy machinery required to carry out the bar fabrication. However, A.J. had a number of highly skilled craftsmen and an excellent credit reference at the banks. In view of the magnitude of the operation and its sponsor, A.J. was able to get the funds to set up a warehouse and a fabricating shop on site, purchase machinery, hire new personnel, and generally get well started on the job.

During the first two months of operations the aluminum company made regular deliveries of the rough cast bars to A.J. These were worked into the bus conductor sections as called for by the accompanying specifications and then delivered to the pot-line area. Many small and some large difficulties were encountered by

A.J. in carrying out the fabrications, but these were being overcome by diligent, hard work, and not respecting normal workday hours.

Suddenly, early in the third month, deliveries of cast bar stopped; within another eight days all cast bar on hand had been completed and delivered to the new pot-line. After a week of no deliveries of stock, A.J.'s local manager made inquiries at the aluminum company office there and was told that they would require no further work by A.J. Very much alarmed, the manager contacted Mr. A.J., owner of the shop. Since less than one-tenth of the contractual work had been completed, Mr. A.J. made arrangements for an immediate meeting with the aluminum company executives.

It was a painful meeting. A.J. was told that the bus fabrication was so far behind schedule and the delivered finished bus was, in fact, of such poor quality that they were terminating the contract immediately. A.J. protested that their production was of the required quality and the rate was completely consistent with the pockmarked, distorted cast bar stock they had been getting, which required much extra hand work. However, protests were unavailing and the A.J. operation found itself definitely closed out.

With the cost of the new facility, machinery, and other financial burdens related to the bus fabrication topping $1 million, A.J. had no recourse but to sue the aluminum company for breach of contract. And the law firm he engaged called on me as an expert to assist them since, as a matter of fact, I had been employed formerly for six years as director of electrical research for a large aluminum company and was very familiar with aluminum production practices.

It soon became evident that the one real issue in this controversy was the character of the raw cast bars as delivered to A.J. for further fabrication. Were these bars straight, well-cast bars as the aluminum company claimed or were they misshapen, with many casting defects, as A.J. claimed? The minor issue was the quality of A.J.'s completed work.

Obviously, my first and most important request was to examine the aluminum company's bar yard and the finished bus being mounted in the new pot-lines. But this is precisely what the aluminum company refused to permit. They cited all manner of excuses, from unwarranted interference with their heavy production schedule to some vague references to classified work involving these areas. Two months dragged by and it began to appear that A.J. might have to drop the litigation, since its costs continued to mount, with no relief near.

Finally, by some magic on the part of our attorney, the aluminum company was served with a court order requiring them to permit our inspection of the bars and buses. Well, they did their best to make a fiasco of it. I was accompanied by two company officials who hustled me to a barbed-wire fenced-in area where cast bars were piled, well back in the enclosure. When I suggested entering the yard, they demurred, saying it was not openable at this time. When I suggested we walk all around the enclosure, they said there were many obstructions. Nevertheless, I then did with some difficulty circumnavigate the fence to where the bar piles were much closer and finally got a good look at their condition. Also, there was some unnecessary confusion about whether the new bus being mounted in the pot-lines was or was not of A.J.'s workmanship. However, I knew where to find the small AJ stamp on the bottom edge of each of his buses and so could check their quality.

My report to A.J. and his attorney confirmed his contention. Most of the cast bars I saw were twisted, some rather badly, and all had very rough surfaces, indicating poor casting control. I found A.J.'s finished bus sections of reasonable quality, no better or worse than the usual practice in most modern pot-lines.

The trial took place in a large town near the site of the aluminum plant. Our attorney called on half a dozen witnesses, all of whom had in some way handled delivery of the cast bars. Then there were another four or five who worked in the A.J. fabrication shop. Finally I was put on oath and requested to tell the court of my examinations and inspections at the aluminum company's plant.

The picture that emerged from these witnesses was that A.J. had been coping with very poor delivered material and only by exercise of great skill and long hours to the job were able to produce a satisfactory finished product. The aluminum company used no outside experts but relied on just a few of its own officials as witnesses to refute the plaintiff's contentions. Their main insistence was on A.J.'s slow production rate and poor quality—none of which was backed up by corroborative evidence, in my opinion.

An amusing situation arose during my cross-examination by the defendant's attorney. He had observed me reading rather hesitantly from my notes and jumped to the conclusion that my eyesight was poor. Suddenly he turned and pointed to a line of engraved dedicatory printing along an upper side of one of the courtroom walls and asked me to read it. I looked at it and said it was too far away for me to make out the letters clearly. At that he quickly turned to the jury and declaimed, "And with your poor eyesight, how can you tell the court and jury that you were able to make the detailed examination on our cast bars you have stated here? Isn't it a fact that you could not see them properly?" I answered, "By no means," as I took out my glasses, put them on, and read aloud every word on the engraved wall printing. The judge smiled broadly, the jury tittered, spectators laughed, and the attorney turned red.

The trial as such was very short. After just two days the case went to the jury. They deliberated for three hours and then brought in a verdict for the plaintiff. A.J. was awarded approximately the full amount of damages for which he had asked. As for me, although I dislike wearing eyeglasses, they did save the day and I have regarded them more kindly since that time.

The Case of the Falling Fuse Clip

This is not a happy story, but there are some lessons to be learned, and for that reason I shall tell it.

The setting was a great alfalfa ranch on the Columbia River. These huge farmlands in what is known as the Columbia Basin involve thousands of acres that must be irrigated continuously when in production; rainfall is extremely light here.

Irrigation is carried out by large rotating arms, some almost a thousand feet long, slowly moving on wheels about a central pivot post. Pumped water enters the irrigation arm at the pivot and is distributed to multitudes of sprayers spaced out along the arm. As your plane flies over such irrigated fields, they present an interesting mosaic of bright green polka dots.

Imagine a five-hundred-acre alfalfa field with ten of these rotating arms each throwing water over an area of fifty acres. Since the arm's wheels are powered by electric motors and the enormous amount of water required is pumped by electric-motor-driven pumps, the electric power demand by that field is very large. Multiply this amount by ten for a five-thousand-acre field, and it is apparent that a major electric supply facility must be dedicated to this industry.

To supply power for the particular field we will speak of the county utility built a good-sized substation within an easement on the ranch property very close to its pumping station. From the substation underground cables brought the power into the ranch's electrical control room at forty-six-hundred volts, and from there service cables went on to the pumping plant to supply its large motors.

Of special interest for this story is the electrical control room, a small concrete building with no openings but one steel door. Inside in steel cubicles were the switches, overload circuit breaker, and instrument meters to control and monitor the forty-six-hundred-volt power flow to the irrigation complex. This control room had been involved in a bad accident several years before the incident that brought me in contact with it, but I believe there were some implications of the one on the other.

The earlier mishap occurred when the ranch owner asked the manufacturer of the circuit breaker system to reset their overload and ground relay operating limits for closer tolerances. A young engineer was sent out to do this job. To work on the breaker relays he knew he had to deenergize all the equipment in the control room. There were two options available to do this. He could request the utility to open the circuit in their substation, or he could simply open the incoming line switch in the control room. To open the substation line he would have to file a request at the utility's dispatcher's office. Since that could take some time, he decided to use the quicker method of simply opening the control room switch. So, after tripping the circuit breaker to insure the absence of any loading, he opened the incoming switch cubicle and pulled the switch's operating handle to the wide open position. So far, so good, but then he did something that killed him.

A common practice after deenergizing electrical equipment in order to work on it is to ground it—thus ensuring that there can be no inadvertent reenergizing and any residual charges left will dissipate. Our young engineer had tied a bare copper wire to a ground stud at the cubicle, and with the other end in his hand, after he had opened the switch, he stooped to wrap the wire around the lower, supposedly dead, stud on the switch. The moment the copper wire touched the stud there was a tremendous arc and the man was killed. Reason: this switch had been wired incorrectly; the incoming line side should have gone to the upper jaws of the switch and the load to the lower studs—which are on the movable blade side of the switch. But the line and load side positions were reversed and the engineer had directly grounded the incoming forty-six-hundred-volt energized line.

The significance of this mishap related to the later accident in which I became involved was that the control building equipment had been installed and wired by a local electrical shop and it was an electrician from this same shop who was fatally injured in the second accident.

In the two-year interval after the first accident, the incoming line switch that was totally destroyed was replaced, correctly this time, and some new metering instruments for current and voltage readout were installed. Let me call particular attention to the voltage metering set-up. The voltmeter itself is mounted on the panel of the cubicle, where it can be read easily. Such instruments give a full scale reading at a comparatively low voltage—not over one hundred volts. Therefore, to be used on a higher voltage system, for example forty-six hundred volts, a step-down transformer must be inserted reducing the four thousand six hundred volts to the instrument's much lower voltage for maximum reading. And the voltmeter is calibrated, using the transformer ratio, as though its readings were the actual line voltages. To accomplish this with accuracy the step-down transformer, known as a potential transformer, or PT, must be a precision device with its transformation ratio stable and known to every close tolerance.

As mentioned above, such a PT voltage metering unit was installed in the ranch's control room. Since the forty-six-hundred-volt power source was three-phase, a three potential transformer unit was used, one transformer in each phase, with each phase having its own cartridge fuse set in the usual fuse clips located between transformers.

On the morning of the accident, the local electrical shop (the same one) had sent three electricians out to the ranch to do some repair work on a pump motor. These fellows were familiar with the electrical system, and before going down to the pumping station they rolled out the circuit breaker in the control building. This circuit breaker was a roll-out type, meaning that when the steel cubicle door was unlocked, the whole circuit breaker could be rolled out and by so doing all the following electrical equipment and cable, et cetera, were separated or disconnected from the supply line. And as long as the breaker remained rolled out there could be no way to reenergize the ranch's electrical system.

It was approaching noon when the electricians completed their work in the pumping station and started back up to the control building to roll in the breaker so as to put the irrigation system back into operation. As they entered the building, the foreman remembered he had brought with him some new fuses for the potential transformers. So they left the breaker rolled out while the foreman opened the potential transformer cubicle and prepared to change its fuses. He wore no rubber safety gloves, nor did he take any other precautions. For safety he relied on two factors: (1) Opening the P.T. cubicle rotated the trunniun plate, upon which the P.T.'s were bolted, outward and upward, automatically breaking connection with this incoming forty-six-hundred-volt line studs; and (2) he knew the circuit breaker was still rolled out and everything beyond that would have to be dead. But he was wrong about that last item, and he should have known better. Voltage metering circuits are always connected directly to the incoming supply line so that the live voltage can always be read regardless whether control switches or breakers are open or closed. And so there were live forty-six-hundred-volt elements in the cubicle. Well, he pulled the old fuses out one by one, barehanded, and started to put in the new fuses. It takes quite a bit of pressure to force open the holding clip pairs when inserting fuses, and the foreman was hanging over the P.T.'s to get more pressure. He had inserted two fuses into their clip pairs and started on the third, seating it into the inner clip satisfactorily but having trouble with the outer clip. Finally, as he put his weight on the fuse, the outer clip gave way, broke off its holder shelf, and together with its long copper stab connection clattered down into the cubicle base where it fell across open terminals of the forty-six-hundred-volt line, short-circuiting them.

There was a terrific explosion, and massive arcing started between the shorted terminals and grounded parts. In a fraction of a second the arcs engulfed the foreman, burning him severely. His companions, who were only slightly burned, rushed him to the hospital, twenty miles away, but there they could do nothing for him. His burns were so extensive that he never regained consciousness and died that night.

Well, the foreman had been a young man and had a young

family with three small children. Of course they needed help, and several law firms offered to prosecute a suit. The selected firm immediately entered suit against the ranch owners and the manufacturer of the potential transformers.

I will not go into the many complex ramifications that came to light during the course of the several investigations by the attorneys and experts for both sides. In the final analysis the plaintiff's case boiled down to:

Against the P.T. manufacturer: The P.T. transformer unit had factory built-in hairline cracks in the hard plastic shelf to which the fuse clips were mounted. It was such a crack under the fuse clip that caused it to break off and drop into the energized terminals at the bottom of the cubicle. This shelf they maintained was made of a thermosetting plastic that could develop cracks if not cooled properly after casting into its final form.

Against the ranch owners: The building used for the ranch's electrical control was too small to provide sufficient clearances between equipment cubicles to work safely. Also, its one door did not provide a rapid exiting of personnel in case of an electrical accident.

The real thrust of the legal action was against the P.T. manufacturer, and this became more evident when it was learned that the alfalfa ranch was near bankruptcy.

At about this stage I was asked by the P.T. maker to investigate the background and circumstances of the accident and to act as their expert witness if the matter came to trial. I found that:

- The deceased electrician had violated every safety rule set up by national codes for working in a high voltage circuit.
- The three-phase P.T. unit came from off the shelf of a local electrical distributor. It was then modified by a local engineering firm so that it could be mounted on the rotating trunniun plate in the P.T. cubicle. The modifications included replacing the brass bolts holding down the outboard fuse clips with steel bolts and attaching them to long copper stabs for the make-break action of the trunniun shelf.

- The P.T. manufacturer in his many years of producing P.T.'s had from time to time changed the resin used for encapsulating the transformers. It is this that also forms the shelf to which the fuse clips are attached. Reasons for these changes were partly improving the product and partly reducing cost. They had had some fine crack lines develop in the older resins during curing and cooling in the past, but their quality control procedures were particularly geared to catch such defects and reject the unit. The current resin, used for the past five years or more, showed no tendency to form cracks.
- Close examination of the P.T. shelf where the fuse clip had broken away did not indicate an old crack present as cause. Rather, the cast-in brass ferrule into which the fuse-clip-holding screw would be seated was found to have rotated slightly due to excessive screw-in force. This was probably done during the P.T. modification when steel screws were substituted for the original brass ones.

From my findings I concluded that the P.T. unit as manufactured and sold for use at the ranch was in normal, good condition, that in its modification to adapt it to a rotating trunniun mount that portion of the shelf holding the outboard fuse clip of one of the P.T.'s was weakened, that the deceased electrician, using his bare hands, not a fuse puller/inserter tool, had applied excessive force in trying to insert the new fuse, and that the deceased electrician should have known that the metering cubicle remained energized even if the circuit breaker was rolled out (open).

These conclusions and others were presented to the plaintiff's attorneys in long and drawn-out deposition takings. Of course as the major factors in causing the accident became more and more clarified, the essential roles of the engineering firm that modified the P.T. unit and that of the deceased electrician himself were recognized. Unfortunately, the local engineering firm had gone bankrupt a few years before and was now defunct, with its principals scattered and insolvent.

The matter did not go to court. Both the P.T. manufacturer and the ranch owners contributed to a moderate settlement to close the case.

What can one say? Both of these fatalities at that alfalfa ranch were so unnecessary. Bad workmanship, poor or no state inspection, and, most of all, extremely poor judgment were the guilty elements here.

The Case of the Distressed Professionals

As everyone knew at the time, the first artificial satellites, *Sputnik I* and *II*, were successfully launched into space by the Soviet Union on October 4 and November 3 in 1957. What followed is also well known—a political/scientific race between the U.S. and the Soviet Union to achieve new firsts in the exploration and perhaps control of space.

By 1959 the U.S. was pouring billions of dollars into its new space program. By 1966 we had achieved a regular manned orbiting of the moon and some nonmanned soft landings. And in July 1969 our *Apollo II* spacecraft put human beings, Americans, on the moon for the first time.

During this same period when these earthbound barriers were being shattered by science, the military units of technically advanced nations were also spending billions developing long-range rocket missiles. We were no exception.

Then, beginning in early 1970 in this country, the space and rocketry programs dried up with startling abruptness. Federal appropriations to the many small and large companies doing research, design, and building of space components were drastically reduced. For thousands of workers in the space industry 1970 became a "black" year. Layoffs were the order of the day, not only for hourly and skilled workers, but right up through the top echelons of professional engineers, scientists, and technical management. By the middle of 1972, the industry was at its lowest point. Many engineers and scientists who had commanded high salaries in their dedicated specialties had by now gone through their financial resources and were on relief or working as casual laborers.

As the situation worsened, the administration suggested that since other areas of American business were in excellent condition, it might be possible to retrain these highly educated and skilled professionals to fit into some of those ongoing industries. The suggestion became a reality when the Department of Commerce announced the government would fund six to eight professional retraining programs in five industry production areas. The grants would go to universities that could show an outstanding capability to provide the specialized education required for any one of the retraining areas. As I recall, the program areas selected were finance, management, transportation, and electric power.

Many universities responded to the grants proposal and filed applications for the various retraining areas, listing their facilities and personnel as especially excelling in the programs selected. One of these universities was the University of California at Los Angeles, a large and well-known university situated in Southern California. When the Department of Commerce first announced the retraining grants, this university called me and asked if I would consider assisting them in applying for a program in electric power and, if they were successful, direct the program there. Their offer appealed to me, and I accepted. I had just retired from Washington State University a few months before after fifteen years in electrical engineering teaching and administration. I thought the possibility of retraining and starting new, rewarding careers for the displaced professionals was a most worthy goal, and I could and would try to make it work.

UCLA did win a grant for retraining in electric power, and in the spring of 1973 my wife and I moved down to Westwood, close to the university, to get the program under way. By the terms of the grant, we were to start approximately on July 1 and carry through to September 1. Within that period it was expected that our very knowledgeable students could assimilate a well–laid-out and presented condensed curricula of power generation, transmission, distribution, and utilization and thereafter find comparable positions of technical responsibility in the electric power industry.

Also, the grant specified that there was to be a limit of fifty persons in each program and that the Department of Commerce would make the student selections based upon applications submitted to the department by those seeking new careers.

I had about three weeks to put together a curriculum, select a staff, organize classroom schedules, and set up the necessary logistics and class attainment criteria and goals. I must say UCLA's staff in their Continuing Education (Extension) Division was of tremendous help in smoothing out the many sacred traditional obstructions and regulations we encountered.

My staff was an excellent cross-section of power engineering professors recruited from UCLA and two or three other California universities near Los Angeles. Between us we worked out a concentrated, embracive study course hitting at every important facet of electric power. This was certainly not a simplified version, quite the contrary, and we kept our fingers crossed that our students, many of them brilliant technical people, could cope with it.

I will never forget that morning when we, the new instruction staff, the UCLA Continuing Education officials, and the fifty preselected student group we were meeting for the first time came together. What a fine-looking, intelligent group of people! They ranged in age from about thirty-five to sixty and included three women. More than half of them had Ph.D.'s in various scientific disciplines, and all of them had many years' experience in their individual specialties.

The director of continuing education gave the greetings and told something about the national program of new career training just starting at some dozen universities. He then turned to me to describe the UCLA program.

I introduced our staff and told of their education and professional backgrounds and for what power specialty each would be responsible. I told the students how much we looked forward to this opportunity to work together to shape new careers and make possible a return to good-paying, interesting jobs. I also told them they could expect a very concentrated, rigorous classroom schedule every weekday from 8:00 A.M. to 5:00 P.M. and a half-day laboratory session on Saturdays—all of which would continue for eight weeks. I explained that the grant provided for each enrollee a small but sustaining weekly salary and additional compensation for transportation costs. And finally I assured them that UCLA recognized that we were dealing with mature professional people, not young students, and they would be treated as such, with cordiality and respect.

Many questions were asked, most of which disclosed some skepticism about the program's chances for success. Could eight weeks fit them for jobs in a completely different field from their training and experience? Some definitely thought not; others were glum about dedicating themselves to two months of hard "back to school" work, still with an unforeseeable future.

The upshot of the meeting was that thirty-nine elected to enter the program, the three women physicists being amongst these, and we spent the next few days processing their entrance forms and arranging for financial compensation.

Classes started during the last week of June, and from then on we rolled along continuously to September. Through my consulting contacts I arranged to have notable people in power engineering and management come in and provide a lecture each week on electrical equipment manufacture and testing and on utility practices. These, coupled with several field trips to major transmission, substation, and generating sites, provided relief from the classroom grind and helped instill additional enthusiasm for the new careers in the making.

From the day I agreed to direct the returning program I was quite concerned about jobs for our eight-week graduates. It seemed to me that one of the keys for helping with this problem might be the right kind of publicity. Therefore, with the assistance of UCLA's public relations people, we ran several news stories about the grant, what it was intended to do, and the interesting backgrounds of the UCLA participants. As a result I received inquiries from several local manufacturers of power equipment.

This gave me the idea that individual manufacturers and

utilities might be persuaded to sponsor one or two of the retrainees, with the ultimate object of providing an employment opportunity. Following up on this, I contacted either the chief engineer or the V.P. of engineering of about thirty companies.

This turned out to be a successful approach. Within three weeks of our start, more than half of the students were accepted for sponsorship by half a dozen utilities, some consulting firms, and several manufacturers. Also, representatives from most of the companies visited the program and met the individuals they were sponsoring. We were very happy about this development, more particularly since it was not planned by the granting agency.

It didn't take us long to realize that fears that our students might not be able to cope with the course's hardships were groundless. The ease and accuracy by which they assimilated what were certainly new physical and mathematical concepts was amazing. And their searching questions kept the staff very busy reviewing reference fundamentals. In other words, keeping up with the class was as much a chore for the teaching staff as the work was for these students. But it was a very satisfactory experience all around.

As the last of the retraining weeks approached, it was most heartening to see how enthusiastic and confident in their new knowledge the class had become. For one reason or another, seven students had dropped out by the end of the first month, but the remaining thirty-two went on through to the end. Those not committed to a sponsor were sending résumés and applications for employment to companies from a list of hopefuls we had prepared.

Graduation day was a rather gala occasion. What a change the last two months had made. Our students looked and acted like professionals again. Twenty-three had responsible jobs awaiting them, several with east coast utilities. The others were not committed yet but had very hopeful prospects.

As for me, the whole affair was a very rewarding and happy experience. My wife and I enjoyed these weeks living in the Westwood area. We had a number of good friends living in Los Angeles, also chamber music players, as we are, and we spent many delightful evenings playing together.

The Case When the Off Was On

Countless times we have flipped a wall switch or pushed a button or turned a switch knob to shut off an appliance or turn out the lights or the TV. Having done this, we never gave the matter another thought. And why should we. The vacuum cleaner was off, the lights were off, and the TV was dark and quiet.

All through commerce and industry a billion big and little switches of all kinds, operated directly, remotely, or in times sequence, are in constant use turning electrical devices on and off. And these devices do turn on and off—as directed by their switch masters.

But now let me tell you about a very large electrically operated "device" whose control switches were all turned off, yet it refused to obey that directive—and there were some dire consequences.

The setting of this case was a major west coast seaport. Here there were many huge grain storage elevators and very large grain ship loading facilities. In a typical operation of the latter, the stored grain is blown through pipes and flexible tubing onto a high traveling boom-crane carrying the delivery chute and this is positioned by the crane operator to discharge the grain directly into the hold of a ship moored against the elevator dock. In this way a hundred tons of grain can be loaded aboard a grain ship in several hours.

It must be appreciated that these loading cranes are very special structures hundreds of feet high, very massive and requiring complex control facilities to coordinate the action of many motors for precisely positioning the crane. In this country such cranes are built by only one or two companies and each is custom designed for a specific location.

Three such cranes were being built by the AB Company for this particular seaport grain elevator facility. Delivery of the first crane, designated no. 1, was marred by an unusual accident. This unit was brought up the river on a special barge equipped with heavy hoisting gear to set the crane into its prepared place on the elevator structure. Just as the crane had been lifted some seventy-five feet, the hoist cable came off the barge's winch drum and the forty-three-ton crane fell into the river, striking the barge on its way. This brand new crane then came to rest in the river bottom mud about thirty feet underwater, and there it lay for almost a week. Eventually the fallen crane was dug out of the mud, lifted onto a barge, and taken to a local sheet-metal fabrication plant for repairs and rehabilitation.

Work on the damaged no. 1 crane appears to have taken only some thirty-six days. During this period the mechanical and electrical repairs were supervised by personnel of the crane's manufacturer and by their consultants, the CD Company.

Later, when I became involved in the case, I tried to find out exactly what was done to put this crane back into working condition. But the repair records were very fragmentary and the few workmen I talked to either had forgotten what was done or made guarded statements. I did learn that besides replacement of the damaged structural members, the motors were dried out and some inconclusive tests were made on the cable wiring, but none of that was replaced.

During this rehabilitation period of work on the no. 1 crane, the other two cranes, nos. 2 and 3, were delivered and installed at the grain elevator without incident. No. 1 crane was then reinstalled, also without incident. With the three cranes in place, the work of wiring them into the computer control post, setting of limit switches, and training of operators began. Within another six months these elements had been accomplished and the cranes went into regular ship loading operations.

Seven years later it happened! For seven years these cranes were busy filling grain ships—mostly from the Orient. Some few troubles had been experienced with the cranes' operations from time to time, but nothing serious. Apparently the grain elevator management had never established any regular maintenance program for the cranes.

On this particular Sunday night there was a thunderstorm, with very heavy rain during the early evening. All the cranes and the elevator itself had been shut down tightly, as usual, since the day before. A partly loaded ship in place under crane no. 1 was also dark, awaiting complete filling of its hold on Monday morning.

At about 11:00 P.M. the boom of crane no. 1 suddenly started to rise. No operator was on duty; no one was in the control room or anywhere else in the entire elevator. The boom rose steadily, soon to a height well beyond its normal required operation. A designed safety stop was built into the crane in case an operator should inattentively permit the boom to rise too far. This was a limit switch set at a maximum safe elevation for the boom of seventy degrees (from the horizontal), which when actuated shut off power to the rise motor. But that night the boom went right on by the limit switch; power was not shut off, and the boom continued to rise toward a vertical position. Before it got up that far, it slammed into the support girders of the elevator structure.This caused part of the structure to cave in and the top part of the boom to jackknife into the tangle of metal—bringing everything to a halt. Well, not quite everything. The heavy jarring and vibration of the boom impact loosened and heaved the big chute trolley out and down onto the grain ship below.

The screech of metal rasping on metal roused the elevator's night watchman, and he gave the alarm to the elevator manager, to the fire department, and to the power company. What they saw when they arrived was a real mess. And what was later estimated to be the costs of repair and replacement of the various damages—to the crane, to the elevator, and to the ship, et cetera—of this mess had a $3 million tag. The elevator company was faced with the immediate task of determining why this crane had self-activated and what or who was responsible for its action. So they brought legal action against each of the companies involved in their acquisition of crane no. 1—its manufacturer and their consultants, the local companies involved in repairing and installing the crane, and all of the agencies that had transported the crane to location at the seaport.

During the next twelve months the many law firms of the plaintiff and the several defendants indeed sought to determine the cause of the accident and the responsible party or parties. The experts hired by these firms, along with engineering and operating personnel from each of the litigants, developed a welter of opinions as to how the crane came to self-activate. Therefore, by the time I was called to look into the case some fifty statements and depositions had already been taken, and from studying these I gathered a general picture of what had been developed as a consensus of the accident. Briefly, this was as follows:

1. At the time of the accident the elevator had no main power switch to shut off all power to the grain loading cranes. The only shut-off for the crane motors was the off-on switch located at each of the crane's operating consoles.
2. Remote control of the crane motors by an operator at this console (located high up in the elevator) was made possible by a twenty-five-conductor main control cable running from the console out on to the crane structure, where it was terminated in a large junction box.
3. This control cable had within its bundle of insulated wires a pair of wires that if shorted together could independently start the crane's boom rise motor and cause it to go up. Also, under that special condition the limit switches would be bypassed and thus be unable to abort the upward travel of the boom. This would occur regardless if the boom switch on the operator's control console was on or off.
4. When the crane's junction box, into which the twenty-five-conductor control cables terminated, was opened, some of the wires' insulation were found to have deep cuts and these showed electrical discharge paths and carbonization,

indicating faulty current flow between them. Most important was the finding that two of the damaged wires were precisely the pair capable of self-activating the boom.

5. The consensus had been reached by the grain elevator attorneys and reluctantly by AB Company, the crane manufacturer, that while installing the twenty-five-conductor control cable the workman had cut too deeply in removing its jacket at the junction box and sliced into some of the underlying insulated wire bundle. This they said was the basic cause of the accident, aided and abetted by an alleged improper-sized junction box bushing, which permitted moisture to enter and eventually short-circuit the damaged critical pair.

One of the reasons the crane manufacturer went along with the statements contained in 5 above was that they had contracted the wiring of these cranes out to a separate electrical construction company. And so they promptly entered a lawsuit against that company, the E Company, claiming it had committed gross negligence and was completely responsible for the accident. Indeed, each of the original litigants pointed the finger at E Company as *the* culprit and thus saw his way out of sharing any responsibility.

That was when E Company through their attorneys asked me to investigate the accident and determine exactly what had caused it. But things were moving rather quickly by then, and before I was able to finish sifting through the accumulated documents, weigh the conflicting evidence, and work on the crane parts that were involved, the case of the elevator company versus the AB Crane Manufacturing Company, their consultants, et al., came to trial. At its conclusion the jury awarded a substantial judgment favoring the elevator company and divided the damage shares amongst the defendants in accordance with what was considered to be their several responsibilities in the accident. Implicit in the conclusions reached was that the self-activation of the crane was primarily due to cuts in the insulations of a critical pair of wires of the twenty-five-conductor control cable resulting from very poor workmanship.

Of course, after this decision AB Company and its consultant firm, CD Company, pressed their lawsuits vigorously against E

Company, who had wired the crane originally. However, by that time I had become completely informed about the whole matter and all its circumstances and findings. And I was convinced that the so-called insulation cuts, presumed to have been made by E Company workmen, had *nothing to do with the accident.* In fact, the scenario that had been evolved was impossible.

Here are two reasons, amongst others, for the above statement:

1. If there had been deep knife cuts in the insulations involved, the crane, after its sketchy rehabilitation, could never have operated for seven years. Moisture leaks and condensation in the junction box would have short-circuited the cut wires in much less than one year.
2. It was claimed that the insulation cuts were made when cutting into the cables' outer protective jacket to remove an appropriate length preparatory to terminating the twenty-five individual wires in the junction box. However, these so-called insulation cuts as found were not under the cut jacket edge but three or four inches beyond into the terminating section.

My analysis of the true cause of this accident was as follows:

In the first place, the fall into the river experienced by no. 1 crane as it was being installed seven years before was much more damaging to its exposed electric cable system than realized by the agencies charged with repairing and operating the crane. A 43-ton object in a partially restrained fall from 75-foot elevation could strike the water with an impact of some 240 tons. Such an impact must have imposed a severe distorting force on the cables, particularly at any place they were tied down (brackets, junction boxes, et cetera).

In addition, the rain-tight covers on the junction boxes would be heavily jarred so as to loosen their seals, and since the crane lay for five days in the river under a thirty-foot head of water, which means under a water pressure of twelve pounds per square inch, there is little doubt that the junction box filled with sea water and some mud.

The insulation on the conductor of this cable was Buna rubber. This type of insulation absorbs water rapidly, reaching 4 to 6 percent by weight in ninety-six hours' immersion, and undoubtedly such an amount of water penetrated the insulation from the stripped ends in the junction box. Water absorption severely deteriorates the Buna's insulating properties, and it is for this reason that it is not generally recommended for water contact service.

If a proper cable dry-out procedure was not carried out—other than wiping out the water and rust in the junction box and resealing the covers—the individual conductors of the control cable probably dried out slowly after the crane was installed but a certain amount of salt from the brackish river was left on the insulation surfaces.

It is not clear to what extent the repair agency hired by CD Company inspected or tested or cleaned up, et cetera, the 25/c cable and its junction box. Actually, a dry three-sixty-fourths of an inch of Buna rubber insulation should show a megger (insulation resistance measuring instrument) reading of at least 500 million ohms—one thousand feet to ground and twice that between conductors. But a megger test is not enough to show up mechanically damaged but dry insulation wear. To check on such damages a high voltage test is required. None was made.

I am almost certain that the fall was responsible for significant damage to the insulation of these conductors, which later got into trouble electrically. The damages occurred in the throat of the bushing at the inside edge of the rubber grommet. That is where the cable wires were restrained, so that the fall's impact on the cable as a whole would tend to tear the rather soft insulation there—and it would tear it only on one side, as found. Under *dry conditions* at the control voltages involved (480 V. three phase, 277 V. to ground), the wires, even with such damages, could operate satisfactorily for a long time. However, moisture entrance into the junction box by very small, slow leaks at the bushing and certainly by intrained condensation would provide some slight discharge paths (especially if the old salt film is still present) between the damaged insulations that are adjacent to one another and at differing potentials. During the seven years of operation, this produced progressive electrical tracking discharges both to ground and from wire to wire. And with this continuing, the insulation carbonized and eroded, leaving gaps in the insulation. The tracking discharges and later small arcing produced ozone (triatomic oxygen) in its vicinity, and this attacked the Buna, since it is not an ozone-resisting compound, and caused the well-known *sharp* crosscracks in the adjacent insulation. Such ozone cracks are to be found on a number of the wire insulations—all adjacent to the major damage areas—but none of the ozone cracks showed tracking or other discharge carbonization. The well-known action of ozone on rubber is chemical in nature, in which a clean scission of the polymer chain occurs—often appearing like a razor blade cut.

Therefore, there never were any workman-made cuts into the cable wire insulations. There were damages, tears resulting from the fall. And when these started to track and discharge, the ozone produced the cracks, which looked like cuts to an inexperienced eye.

My reconstruction of the accident was as follows:

During the seven years crane no. 1 operated without regular inspection or maintenance of the control wiring, the conductors of the 25/c cable in junction box 1 remained sufficiently dry so that the original fall damages to the insulation did not prevent the boom from operating normally. However, during this time of the seaport's well-known high humidity and heavy rainfalls, periodically some moisture was entrained in junction box 1 and on occasion some of this moisture was drawn into the already damaged areas of the insulation so that small conductive paths were established for a short time across these wires. This produced some electrical discharges between the damaged conductors and to ground—all of which contributed further to insulation damage.

The *self-activation of the no. 1 boom* on the Sunday night came about as a result of some additional moisture establishing a more solid conductive path between two critical, damaged wires. Or this occurred in combination with a transient on the facility's electric supply system (lightning-induced), which came through the control circuit and triggered a sufficiently heavy discharge to constitute a short circuit between the critical conductors. Since the control circuit design permitted a short circuit across these wire pairs to directly activate the boom's raise motor, at the same time bypassing the limit switch, the accident happened.

There were many contributing factors to this accident. The

evidence indicates that some electrical discharges had been going on for some time between one or more of the damaged conductors in the control cable and to the grounded portions of the cable entrance bushing in the junction box.

The control circuit was protected for ground faults by low rating fuses and a ground relay operating a contactor to clear any phase-to-ground voltage involved. Since ground faulting cannot be tolerated in a control circuit, these protective devices are required to be designed and selected to operate on very small ground currents—at not more than one ampere of current flow.

But it appears that during the long period in which ground faulting in 25/c cable was going on the protective system did not operate. Had it operated properly, the accident would not have occurred.

The grain elevator facility and the crane designers provided no practical, ready means for the operators or maintenance personnel to shut off all power to the cranes at night or over weekend periods. Actually, if the electric system had been designed with a proper shunt trip breaker and an automatic timer, all power to the cranes could have been shut down and reenergized automatically at preselected times. With this type of automatic energy control the accident could not have happened.

Also, the control system design utilizes three-phase A.C. voltages at 480/277. This is comparatively high voltage for control circuits. A low voltage (twenty-four to forty-eight volts) D.C. control system would have operated satisfactorily even though the insulation on some of the cable conductors was damaged, as these were.

Also in the design of the crane control system there should not have been a critical pair of adjacent wires under the same cable jacket whose short circuit could directly activate the boom, leaving the limit switch bypassed. This is also true of the open face terminal board mounted in the junction box, since a short circuit by moisture condensing across the right terminals there would do the same thing.

I presented my findings in a long position paper to the attorney representing E Company. He agreed that the evidence of E Company's negligence as brought out in the first trial was in error and that we could prove E Company had no responsibility in what had happened. To reenforce this I prepared an affidavit stating my convictions regarding the accident cause, and this was presented to the judge who had heard the first trial.

At this writing, then, the matter still rests. The mills of jurisprudence can grind very slowly indeed.

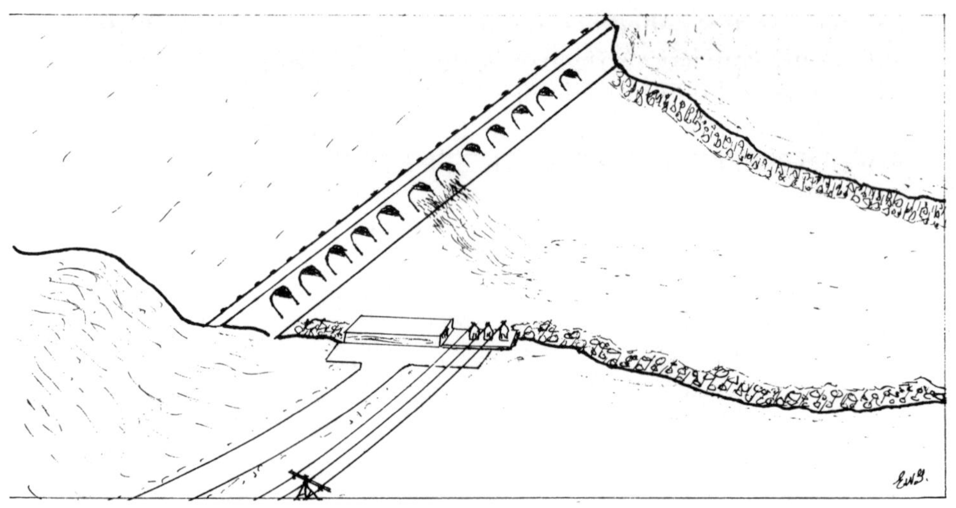

The Case of the Dam Dilemma

Horace Greeley's oft quoted axiom "Go west, young man" has always implied "and you will find what you need." Indeed, to "go where it is" was a primeval urgency practiced by the animal world as well as mankind. All animal life moves with the seasons to better climates and food resources. And primitive man, as well as our more recent Indian and nomad cultures, did the same.

With the settlement of communities of men and their growth into larger and larger built-up environments, the "go where it is" expedient became virtually impossible. The city's structure is unmovable, and its inhabitants' lives become interwoven in that structure. Whatever climate prevailed must be endured year round, and whatever food and water is needed must be brought in. And, since modern cities need energy in various forms, this too must be brought in.

Confining our attention to energy needs and particularly to electrical energy, it is indeed a fortunate city that is located near a ready low-cost source of electrical energy. That is the situation, the fortunate situation, of the cities and towns in our Northwest. The rivers of the Northwest, running from the great snow fields and ice packs of the far north, have provided a continuing great and ready source of electrical energy for the entire region.

Everywhere on these rivers of the Northwest there are dams for hydroelectric generation and also for irrigation and flood control. Generation of electricity by falling water is still the most economical of all power producing methods and has the least disturbing effect on environment quality. Capitalizing on these virtues, many energy intensive industries have built new plants close to these energy resources and by so doing have caused a boom in the economy of the region. Lucky cities, lucky region!

But progress in our modern American civilization is a greedy consumer of energy. A historical formula that held up until the sixties was that power requirements doubled every decade—which in the Northwest economy meant that more new dams had to be built and enlarged. My story here has to do with one of the old hydroelectric dams that was to be enlarged to double its capacity but there were some unusual problems.

This was a modest-sized hydroplant on a modest-sized river near the Canadian border. For thirty years it had served the needs of a large coastal city, supplying it power in conjunction with several other smaller dams. Now the city with its outflung residential suburbs and new industry expansion needed more power and foresaw that its need could continue to grow beyond the capability of the generating plants currently committed. Their largest plant was located on a river whose flow could support new generating units equal or greater to what was in place, and so they decided to do just that.

Like most utilities do, the design and construction of the generating facilities were let to several consulting engineering firms, but the selection of equipment and materials was firmly in the hands of the city. Bids for the generators, auxiliaries, new power house, dam extensions, and associated piping and gear were issued by city engineers with a low bid as mandatory purchase requirement.

A foreign manufacturer (Oriental) provided the low bid on the three generators and was accepted. As the purchase specification was finally adopted, it called for the shipment of all generator components and parts to the dam site. There the components were to be erected, interconnected, and proof tested by the purchaser, using a complete set of instructions supplied by the manufacturer. However, manufacturer's representatives were to be in residence at the dam and would act as consultants and inspectors of the work being erected.

These high capacity generators driven by relatively slowly revolving water turbines are very large. The outer fixed circular segment, known as the stator, is almost forty feet in diameter while the inner rotating so-called rotor spans some twenty-five feet. When the rotor is in place, its outer circumference clears the inner circumference of the stator by less than half an inch. The fixed stator contains the wire coils that, as the strong magnetic field of the rotor sweeps by, induces voltage upon and causes current to flow through them, each coil adding to the total. Thus at a given rotational speed and a given magnetic field strength and with a given number of interconnected stator coils, the generator will provide electric power at a constant voltage and a maximum amount of current—both of which establish its power rating.

Since soft iron offers the least resistance to magnetic fields, both the rotor and the stator are mainly of that material—laminated sheet-metal in split form to further reduce magnetic losses. Also, in a large generator the stator coils are made of formed rectangular copper bars rather than round wire and these are set into slots in the stator metal, flush with its inner surface so as to be as close to the moving rotor as possible. These stator bars, since they have induced voltage and carry current, must be insulated appropriately from their iron slots. Also, this insulation has to be strong enough to remain intact during forming of the bars, wedging them into the stator slots, and then withstanding the electrical and mechanical stresses imposed during the generator's expected long service life.

One early morning during the start of the real winter weather in the Northwest, I was called by an engineer representative of the consulting firm engaged in the dam's enlargement construction. They were in the midst of installing the copper bars into the some 290 slots of the first generator stator. As they completed a group of twenty bar insertions, they would connect them with temporary wire and test the group for insulation integrity. In this test a specified voltage is applied for a specified time between the bars and the stator iron. What they were finding thus far, after installing five groups, was that each group failed under test and then showed very low resistance to the stator iron. When they tested the insulation on the individual bars, 30 percent of them were shorted to the iron slots. The representative said they had been following the detailed instructions furnished them by the manufacturer carefully and were now at loss to understand the insulation failures. I asked if the manufacturer's inspectors were on the job. The representative replied that they were there but had been of little help in this. Could I come up to the dam site as soon as possible?

Of course I was interested—big hydro generator problems were relatively uncommon. But weather conditions east of the Cascades had turned rugged, with heavy snow toward the Canadian border. Also, there was no public transportation of any kind up to the dam's remote area. When I mentioned the bad travel conditions, he immediately offered to send down a heavy four-wheel-drive truck to pick me up. So I went up the 120 miles to the dam on snow-filled roads in safety if not in comfort.

The river and dam setting amidst evergreen forested mountains banked with snow made a breathtaking picture. But there was no time to explore its beauties. I was hustled into the engineer's portable office to meet the construction staff and then taken down to the partly completed new power house where the generators were being assembled. During a few hours talking to the assembled electricians and the inspectors and examining the preformed copper stator bars not yet installed in the stator slots, I learned several important things. First, the shielded insulation covering the copper bars was glassy hard, with no "give" whatsoever. Second, and equally important, the inspectors spoke very little English and understood almost none. With much gesticulation and voluble speech they tried to tell me what was the matter with the bar installation. But I could understand little, and of course that was also the case for our local installing crew. So it was better for me to find out for myself.

Entry into the inner circle of the stator where the bar installation was going on was by several very low tunnels through its seven-foot width. Crawling on hands, feet, and belly, I finally got into the inner stator circle, after taking a few good bumps, fortunately on my hard hat. Once inside I was able to observe a crew setting in a group of bar coils. Generally the six-foot-long bars were laid into the appropriate slots and then tapped into place with a rubber mallet. Slight nonuniformity in the metal slots and some inevitable waveness in the coil bars had made the tapping in difficult. At places where a bar would not enter the slot, a soft wooden wedge was driven in between the bar and the slot side and then the mallet applied hard enough to drive the bar into the slot while the wedge was being removed. This procedure left the bar tightly against the slot wall in places but with appreciable gaps in other places. They had been filling the gap regions by driving in different thicknesses of rigid plastic shims that were supposed to be conductive.

Let me pause here to explain something about these stator bar coils. Generators of this size and power rating are built with much higher output voltage than older machines. This means that stator coil bars must have comparably higher strength insulation than formerly. The bar insulation is in more or less contact with its metal slot—which is grounded. The thin conductive covering over the bar insulation, known as the shield, is there to ensure that all of the bar's insulation is in intimate electrical contact with the slot walls. In the absence of a shield covering, air gaps between insulation and slot wall could break down, spark over, and then arc—which would destroy the insulation in short order.

I have already mentioned that I found the insulated surface of the coil bars very hard. On closer examination and dissection of some of the bars, it appeared that both the insulating material and its outer conductive shield were "rock-hard." Also, the conductance of the shield on many of the bars was relatively poor. The suspicion was forming in my mind that where a bar had to be wedged and hard-hammered into the slot the glasslike shield might be cracked and provide openings for tiny air gaps. These air gaps would discharge under the test voltage and thus cause the insulation to fail. That idea set me to examining a number of the coils that had failed when tested in place.

All of that shift electricians, the construction engineers, and the inspectors gathered around me as we laid out two bars that had failed on a rough table. We had three or four seven-power magnifying glasses to hand around, and I had on a seven-power binocular headpiece. In each bar the actual failure place was easily seen—where the arcing had carbonized the puncture through the insulation. But in addition we found numerous cracks in the shield at every location where it was obvious a wedge had been used to pry the bar into place. Some of these cracks showed brownish electrical discharge stains; others did not. In one case a crack went through the entire insulation down to the copper but no failure had occurred.

So here was the answer to the coil failures: the hard, brittle nature of the bar insulation material and the method of forcing the bars into the stator slots. We looked at the rest of the new coil bars that were stocked in the power house intended for this first generator. All were similar to those already inserted into the stator. All had the same hard, glassy shield/insulation covering. We could therefore expect that unless the method of inserting them into the slots was changed drastically in some way, none could avoid the possibility of failing under the voltage test.

By this time the inspectors were shaking their heads in unison with ours. The managing engineer of the project put in an overseas call to the generator manufacturer to explain and to complain. He was at this for nearly two hours. When he finally came out of his office, he informed me that they would stop work with the present coil bars and segregate them all together for ultimate shipment back to the manufacturer. Also, the manufacturer's engineer had told him that the coil bars for generator no. 2, now stocked in the power house, had a different type of insulation that had more resilience and he suggested that these should be used now in the first generator's stator. Also an additional inspector, English-speaking, would be flying to the dam site immediately, bringing new tools and methods for inserting the bars without harm.

I looked at the no. 2 coil bars and found that the insulation covering was hard but not glassy. Perhaps it did have some slight resiliency; the other bars had none. At any rate, I left at that point, since it would be several weeks before work on the stator with new coils and new insertion methods could be started.

On my next visit to the dam, my wife and I drove up. The snow was off the road, and everything was dripping in the warm sunlight. With all the melted snow swelling the river, the great mass of water coming over the dam face was fantastically beautiful.

In the power house I found about half of the new bar coils already installed in the stator. The voltage test results on these were not perfect, but were much better than with the first set of coils. Fewer than one in ten of the bars were failing the test, whereas formerly between three and four in ten had failed. The construction engineers on the project said they could live with that failure rate and ascribed the improvement mostly to better insertion wedges and elimination of the rubber mallet entirely. I also checked the conduction of the insulation shield on several bars and found it appreciably better (low in resistance) than the glassy shield on the old bars.

Before I left the dam, I told the project managing engineer that I was not satisfied with these coil bars, although they were an improvement. The windings of the generator in operation are subjected to a great deal of vibration and temperature changes. In particular, the stator coils are in the strong alternating electromagnetic fields and the pulsing windage of the moving rotor plus its mechanical vibration. Experience in this country and others has shown that a stator bar insulation needs to be highly resilient while in snug contact everywhere along its slot to withstand these vibrational stresses over a long service life. I told the manager that I hoped it wouldn't happen but suspected these stator coils might start failing after four or five years of service.

At this writing the new generators have not yet been placed on the line. I'll keep my fingers crossed and wait for time to tell.

The Case of the Dreadful Dredge

Anyone familiar with Southern California will readily recall those concentrated areas of greeny-brown earth laced with pipes running between corrugated steel tanks, some small, some large. And in their midst are what appear to be large iron donkeys continuously lifting their blunt heads up and down, their teeth clamped about a steel bar moving rhythmically into and out of the earth. Once oil was discovered under the rich California soil, the rush to get it out was even greater and certainly more lasting than the earlier feverish stampede for precious metals. Wherever there was a showing of oil a quick field was opened. Some of these were very extensive, taking in what were once large citrus orchards, but many were in the backyards of dwellings or incongruously in the midst of a recreation park or even right in the middle of a country road, forcing it to detour. Iron donkeys moving up and down pumping the liquid gold were almost everywhere to be seen along the coasts of Southern California. And this California crude was of exceptionally good quality, greatly in demand.

Today as I write, much of the abundant reservoirs contained in the earlier discovered sands has been depleted and it is no longer profitable to keep the many donkey pumps going. Fewer and fewer fields remain active, but there are still donkeys here and there bobbing their heads in acknowledgment of past glories.

Once it became certain that the flow was dwindling from the established oil fields and even newly opened wells, the search moved offshore into the ocean. And there the searchers "struck gold" again. Less than four miles out in the ocean and under relatively shallow waters, test drillings tapped into immense oil-bearing strata; the rush was on again. For a time it appeared that anchored oil rigs and derrick platforms would proliferate all along the coastline. However, a few foulings of the California beaches brought on strict federal and state regulations limiting the number of permitted ocean well exploitations and closely prescribing their operating procedures.

A short distance south of Los Angeles an off-shore drilling operation was so successful that it became expedient to build a whole artificial island to house the many well heads, storage facilities, operating equipment, and personnel. In addition, as the crude oil filled the island's storage tanks, it was pumped via large submarine oil lines to mainland facilities and treatment centers. To satisfy the oil island's obvious large energy requirements it was necessary to provide the needed electric power by submarine cables fed from an on-shore dedicated substation of the local electric utility. And here my tale of the dreadful dredge unfolds.

In order to supply the oil island's full operation demand and also provide coverage for some future expansions, the power company decided to install two high voltage submarine pipe-type power cables operating at 230,000 volts. These were laid in the channel bed some 500 feet apart and continued for 13,800 feet (a little more than 2½ miles) to the island, where they were terminated. Choice of two separate cables was not only to provide ample capability for future load expansion, but should one cable fail or be put out of commission, the second cable could for a time carry the island's basic energy requirements.

At this point I need to briefly describe what we call a pipe-type cable. It has long been known to safely insulate high voltage across the relatively thin insulation walls of a cable, that insulation, besides having the highest dielectric strength possible, must never permit any unfilled voids, no matter how small, within its structure. Such voids have very low ability to resist voltage stresses and will readily spark over and generate arcing, which soon destroys the insulation. To prevent any possibility of void formation in high voltage cables, the insulation (usually high grade paper tapes impregnated with insulating oil) is kept under oil pressure from an external source.

In a pipe-type cable the insulated conductors are encased in an oversized steel pipe and the whole inside of the pipe is filled with insulating oil kept at pressure of several hundred pounds per square inch. In this way, as the cables inside the pipe expand and contract with load changes, the insulations are kept compressed under a constant high oil pressure—thus preventing formation of any voids.

A pipe-type cable is mostly completed in the field. Forty-foot sections of specially clean, deburred pipe are welded together at the site to the desired length. X-ray checks are made of each weld to ensure tightness. The full-length pipe is laid along the selected route between supply and load terminals. The pipe is then evacuated at low pressure to eliminate moisture and again checked for leaks. Dry nitrogen gas is then introduced as interim protection at that time. Meanwhile the three oil-impregnated insulated conductors, factory made and held together by a wrapping of water-resistant plastic film, are delivered to the site on a large reel with a sealed outer wrapper under which a positive pressure of dry nitrogen gas has been maintained. A steel pulling cable, previously strung through the pipe, is attached to the pulling eye fastened on the end of the three conductors, and with a power winch at the far end the conductors are drawn into the pipe. As the insulated conductors enter the pipe, they are doused with previously dried and outgassed insulating oil. Once the conductors are fully in the pipe and terminated in their sealed enclosures, vacuum is again applied to eliminate any picked-up moisture. After this, the pipe is completely filled with insulating oil at pressure of two hundred pounds per square inch, and this is maintained throughout the cable's life by a pumping plant and large oil reservoirs. By such means it has been possible to operate pipe cables at 375,000 volts with safety and long life. Pipe-type power cables are very popular in this country.

On a Friday afternoon when the oil island was in full operation, a low oil pressure alarm sounded in the power company's dispatcher's office. He pinpointed the trouble at the substation supplying power to the oil island and sent a troubleshooting crew out. When they arrived there they found the pumping plant running continuously and the oil reservoir one-third depleted. Pipeline pressure was down to less than fifty pounds per square inch. Evidently one of the pipelines had sustained a major leak and the pumped reservoir oil was simply running out of the pipe somewhere. Fearing an insulation failure, they opened the circuit breakers to the oil island, killing that pipe cable.

Just about that time the power company received a telephoned message from the local office of a large maritime dredging company. One of their dredge operators working in the channel reported that his spud bucket had struck an object on the channel bottom and oil was coming up to the surface. They therefore presumed that the dredge had struck a pipe cable, since they knew it was laid in the general vicinity. Well, there was no doubt about it—the pipe had been struck and it had been ruptured.

The utility engineers reasoned that since there was no electrical failure, the pumped-in oil must be keeping the water out of the pipe and therefore, if they could keep oil flowing out of the break until the pipe could be repaired, they could probably save the cable. To this end they moved in several large tank trucks filled with insulating oil and connected them up to supply the pipe cable as needed. It was hoped to get a diving bell in place by the next morning and have some expert pipe welders complete repairs by midafternoon. But it didn't work that way. A suitable diving bell and auxiliaries were not available for at least three days. During that interim the reservoirs and truck tanks ran out of oil for some eighteen hours—and sea water pushed into the pipe.

A diving bell eventually was placed over the broken pipe lying on the channel bottom, and under comparatively dry conditions the welders made an excellent sleeve repair. At that time, the utility called in a cable consultant to advise them on what could be done. After studying the log of events, he suggested that the cable could be saved and returned to service. His suggested procedure was to force all the free oil out of the pipe, using pressurized nitrogen gas as a pusher. Then the pipe was to be evacuated at a constant very low pressure for seventy-two hours. Then insulating oil would be pumped into the pipe and the pressure slowly raised to two hundred pounds per square inch. After this, the insulated conductors were to be tested at a moderately high d.c. voltage, and if they held, the pipe cable was to be put back in service.

About this same time, the attorney for the dredging company asked me to look into the matter for them. I too studied the log of events and came to the conclusion that the cable could not be saved. My reasoning, from long experience with this type of cable insulation, was that the entering water had displaced the oil and was now absorbed into the paper capillaries. Also, the displaced oil would act as a slugging agent and effectively prevent the vacuum from reaching the moisture, so that little or no drying out could be accomplished.

My opinion on the above was communicated to the utility engineers. Nevertheless, they went ahead with the cable saving program as outlined by their consultant. As it turned out, it was a long and costly program. Almost everything that could go wrong did. Draining the oil from the pipe took more than a month, and at that the amount coming out was less than 50 percent of the known full pipe quantity. However, they decided at that point to start evacuating the pipe. They were at this nearly two months without ever reaching a desirable low pressure for moisture drying. What was coming out in the pump condensate was mostly oil vapor from the thin insulating oil left in the pipe.

By the end of three months it was fairly obvious that the cable saving program would be a failure. And sure enough, when they applied the d.c. test voltage each insulated conductor failed electrically within the fifteen-minute application period. Now the utility was faced with ordering nearly forty-two thousand feet of new insulated conductors, repreparing the pipe, drawing the new insulated conductors in, and terminating and pressurizing the system. And now, too, the utility entered suit against the dredging company, claiming that the dredge that struck the cable had no business being where it was, since accurate maps of the cables' positions in the channel had been supplied to them before they started work there. Damages requested included not only the new cable and installation costs and lost revenue from the oil island but also the total costs of their efforts to save the old cable.

I was present at the substation when the old cable was pulled out of the pipe. It took two very large bulldozers to get the withdrawal started, but then the conductors came out more or less smoothly. Those of us examining the cable as it emerged frequently stopped the pulling in order to photograph the cables and look at particulars under magnification. As the pull-out approached the region where the pipe had been damaged, the evident signs of water appeared—damp-smelling paper insulation and seawater corrosion of the copper shielding tapes. These signs extended fifteen or twenty feet to either side of a failure puncture. No wonder this cable failed under a modest test. It should have been pulled out months ago.

As the legal sparring between the opposing attorneys developed with depositions, interrogatories, affidavits, et cetera, it began to appear that the map locations of the two pipe type cables given to the dredging company were not as-built locations but planning locations used for permit application to the Coast and Geodetic Survey. The dredging company contended that those pipes were actually at a substantially different position from where they were said to be in their information as supplied by the utility. And so the contentions and countercontentions were joined legally.

For my part I prepared a technical position paper directed toward the illogic of attempting to rehabilitate a wet impregnated paper cable under the circumstances that existed. I maintained it was almost a certainty that any such rehabilitation procedure would fail. And inasmuch as the utility was made aware of this accurate prediction, the cost of that program should definitely not be accountable to the dredging company.

Evidently this position paper was taken into consideration when the parties finally negotiated a settlement out of court. As far as I know, the settlement covered only a portion of the cost of the new cable (due to the location map uncertainty) but included nothing for the old cable rehabilitation or power revenue loss.

The Case of the Hidden Casualty

The United States had just entered World War II and was frantically engaged in setting up munitions manufacture and calling up manpower. Fort Dix in New Jersey was in a tumult of facilities building, training thousands of raw recruits and draftees and assembling vast quantities of war materiel. As can be imagined, none of this was going along smoothly.

From the first day that the fort was activated, the need for a large centrally located water pumping facility became evident. Water supply to the peacetime fort with its limited staff and requirements became hopelessly inadequate as thousands of to-be troops moved in along with hundreds of trucks, troop carriers, and tanks.

Once construction began, the badly needed pumping plant was put together in record time, but during its month of construction, the fort suffered along on what amounted to short water rations. And so with the "turn-key" completion and handover to the army, the brass decided to hold a dedicatory ceremony at which, after appropriate remarks by War Department and army bigwigs, Dix's commander would throw the switch putting the plant into operation.

Here I should indicate that power to operate the plant was to be supplied over a previously installed large, three-conductor lead sheathed, fifteen-thousand-volt underground cable running from a New Jersey Central Power and Light Company's substation about 1½ miles away. Also, and most important to this story, the cable was manufactured by the large wire and cable company where I was employed as a research engineer.

Now back to the dedication. That morning at the new plant's master control panel, the main supply circuit breaker was specially decorated for the occasion and there was a red ribbon leading to its automatic operating button. Pushing in that button would activate the entire plant's electrical system and set the massive pumps working.

With the speeches over and building and forecourt crowded with army brass and invited guests, the commanding officer of the fort, a lieutenant general, stepped up to the control panel and smartly pushed in the breaker button. What happened shouldn't have happened—and, in fact, nothing did happen. No lights came on, no equipment was activated, and the pumps remained silent.

What a scurrying there was then. Eventually checks by the plant's contractors and army engineers confirmed that everything was connected properly and all electrical equipment was in good working order. Also, the supply substation had been set to deliver its power to the plant on demand—but there was no demand. Ergo, the problem must lie in the cable connecting the substation to the plant, and further, this cable must have been totally defective and have failed at once. As soon as this conclusion was reached, our company's chief engineer was peremptorily called down to the fort and given to understand that supplying defective equipment to the army during wartime could be a treasonable and criminal offense, with dire penalties.

I was just sitting down to dinner at home when the call came. "Gene," my chief said, "we're in deep trouble at Fort Dix—our cable to their new pumping plant has failed. Get everything together you'll need for fault locating and get down there p.d.q.—leave now, and that means right now."

So I rushed over to my laboratory, piled everything I could think of into the car, got one of the company's gas ration cards, and was off at about 7:00 P.M. to travel the hundred or so miles to Fort Dix. Remember, at that time the blackout was in effect and some road signs had even been removed. At any rate, I arrived at Fort Dix at about 9:00 P.M., after a hairy but safe drive.

There I was met by a very worried man. He was the local electrical contractor who had been hired by the army to install our cable some months before. He told me that ever since the dedication fiasco he had been digging holes at every likely place along the cable run that he thought might have contributed to the cable's failure but had found nothing.

We spent the next couple of hours walking the line to familiarize me with the lay of the land, and then I started unpacking my instruments. I knew there was something odd about this cable failure. Usually when a cable of this kind fails on being energized, the damaged or defective insulation is punctured by a flashover and a heavy arcing follows, drawing large short-circuit currents, which quickly open the circuit breakers at the supplying substation. But nothing of this kind had occurred, so it looked like this was not an electrical failure we were dealing with. Either there was no cable in place or it was not connected at one end or, unbelievable but still possible, the single splice (joint) in the line had been made improperly and had pulled apart.

Having arrived at this reasoning, we disconnected the cable terminals at both the substation and the pumping plant and I checked for conductor continuity with a resistance meter. *There was none!* None of the three conductors measured continuous between the far terminals. It was hard to believe that all three of these copper conductors, each about one inch thick, could be open somewhere.

To accurately locate such a cable open fault, a capacitance bridge is used to measure the cable's capacity, conductor by conductor, from one end and then this is repeated at the other end. When one knows the design capacitance per foot of cable and has an accurate measure of the cable length it becomes very easy to calculate the exact distance to the fault from either end.

It was close to 2:00 A.M. when I sent my contractor friend along the cable run with his bicycle wheel odometer to get us the actual working cable length between terminals. And it was about 3:00 A.M. when I had completed the capacitance measurements, calculated the distances to the open fault, and located where we would have to dig. The calculations from both ends agreed within two or three feet and came out almost in the middle of what looked like a brand new sixty-foot-wide concrete roadway within the fort. I later learned this road had been opened only a few weeks ago and was now a major link between the fort's vehicular park and the staging areas. You can well believe that in view of this location I spent another half hour checking and rechecking my tests and calculations.

The CO had instructed me to report what I found immediately, regardless of the time, so I called him and with no enthusiasm and no little trepidation told him where I believed the cable was faulted—in the middle of his new road. He hit the ceiling! He shouted at me, "That road's continuous traffic is vital to the operation of our war effort! It's just been built and your locating the trouble under it must be absolutely wrong," et cetera, et cetera. After a while he calmed down and indicated he would have to get clearance from the War Department to break into the road but woe and more woe to me and the company if that turned out to be a futile, destructive exercise.

By that time my confidence in the tests and measurements had literally evaporated, and as you might expect, I got no rest in what was left of the night.

A little after six in the morning, the contractor called me to say he had just obtained permission to barricade and open up the road, which he would start immediately. I got to the site as the jackhammers began to cut into the road. I was quaking in my boots as I watched.

Well, we found it, about fifteen feet into the road. There the cable lay under a number of large broken concrete slabs, and when these were pried off, the mystery was solved. The cable was completely broken in two, with each end mashed and twisted at a sharp angle. Evidently, in building the road a bulldozer or backhoe had caught the cable and ripped it apart. That was bad enough, but in addition, the road crew had concealed the damage and built over it to avoid any responsibility.

When the CO arrived we both stood there looking down at the broken cable—me smiling, him frowning and muttering choice oaths. I wiped my hands, made my farewells, and was off for home—feeling wonderful.

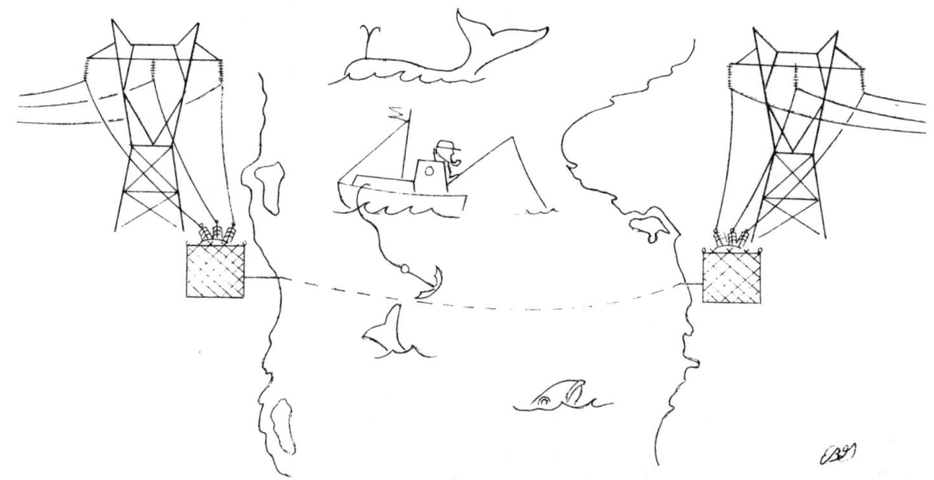

The Case of the Disappearing Armor

Sometime in the early 1970s, a major power company supplying most of the San Juan Island communities projected a power shortage within five years as a result of the burgeoning popularity of the islands for both summer and year-round residents. There are no significant electric power generating facilities on these islands. All power comes from a federal transmission agency of the Northwest, which has access to the many large hydroelectric power plants on the rivers of that area. Consequently, the island utility entered into negotiation with the federal agency and obtained a contractual obligation for the needed additional block of power—to be delivered from a load center already on Lopez Island. With that accomplished, the utility set about securing the necessary permits and the submarine transmission cable to bring the power over to Friday Harbor on San Juan Island, where most of the projected requirement lay. The underwater run was estimated to be about 2.5 miles (thirteen hundred feet).

One of the largest cable manufacturers in the country won the bid and supplied a three-conductor submarine power cable insulated with polyethylene resin. The insulated conductors were bound together with a helically wrapped copper tape, and over a fabric bedding as armoring there was a full coverage of many large-diameter aluminum wires spiraled at a long lay overall. Each of the armor wires had a covering consisting of a thin layer of green polyethylene resin.

The use of covered aluminum wires for armoring a submarine cable was unique. However, the supplier insisted that the aluminum armoring was much lighter in weight and more economical and would last longer than conventional galvanized steel wire armoring, since there could be no rusting.

At any rate, the cable was purchased, laid in the channel, connected up, and energized after only a few months from the time the decision to bring in additional electric power was made. Total cost ran close to $1.5 million.

Eighteen days later this new cable failed in service. Under the island's existing power commitments at that time, the utility had no recourse but to quickly locate the failure, cut out the failed section, and join (splice) in a new short length taken from spare cable that was available. Seven days after the cable was back underwater and reenergized, it failed again.

With that, realizing there was something seriously wrong, the utility informed the manufacturer that their cable was defective and would have to be replaced. In a few days the manufacturer's sales representatives visited the island, looked at the cable failures, and stated that the cable had no defects and therefore they could not accept any responsibility. They claimed the utility had in some way mistreated the cable during the underwater laying operation and this had damaged it and caused the failures.

That was when the island utility asked me to visit the site and look into the cable failures.

When I learned the details of the situation and examined the failure regions, there was little doubt in my mind as to what had occurred. I had a suspicion of what might happen when I was first told about the resin-covered aluminum armor wires. Now that suspicion appeared to be confirmed. Nevertheless, I made a thorough dissection of each cable failure region and was then able to give my conclusions with confidence.

The most significant finding was that in a number of places the aluminum armor wires appeared to have been "eaten" apart and under these places the copper binder tape over the insulated conductors was chemically attacked to an unusual brightness, with some swelling underneath the tape. At the specific failure areas the copper tape had actually split, and the underlying insulation was very swollen around the failure punctures.

With the above telltale signs it was fairly simple to reconstruct the failure events. The thin polyethylene armor wire covering (it was only thirty-thousandths of an inch thick) I found could be easily damaged, even with a fingernail. Obviously, in the laying operation, which is itself somewhat rough on the cable (hence the need for armoring), the armor wire coating received many nicks and tears, exposing the bare aluminum to sea water at those openings. And at such openings a combination of electrolysis and galvanic action between the exposed aluminum and the copper tape, all in a good saline solution, set in immediately. In such a combination the aluminum becomes an anode—that is, it assumes a positive polarity—and the copper becomes a cathode, assuming a negative polarity. As current flows between the anode and the cathode, the metal of the anode goes into solution continuously while the cathode is subjected to a slight attack. The process is increased enormously if the anodic exposure is through a *small* opening, since the electrolysis current is concentrated there. Also, while the process is going on, a great amount of gas is evolved continuously, including mostly hydrogen and metal-carrying gases.

And so the new covered aluminum armor wire proved to be a complete bust. As the armor wire metal disappeared into the sea water at the cover openings, the consequent heavy evolution of gases attacked and destroyed the integrity of the insulation —swelling it out greatly and leading to the electrical failures.

We brought this staightforward analysis of how it happened to the attention of the cable supplier, but they still refused to acknowledge that their aluminum armor design was faulty. Therefore, the utility took legal action.

The controversy went to the U.S. District Court with the manufacturer as defendant. He brought in his chief engineer and his director of research to bolster the claim that their design was sound and had been given all sorts of definitive tests in their laboratory, with no troubles developing. As the witnesses droned on, I was giving much thought as to how we might get across to the jury what actually took place to cause the failures. After a few experiments at home with a couple of mason jars, I had it. I put together a home-made electrolysis cell and prepared to take it to the U.S. District Court for a live demonstration before judge and jury. Our attorneys were delighted with the idea but cautioned that the judge might not permit it in court.

We talked with His Honor, and I explained that the demonstration would enact almost an exact replica of what happened to this cable before it failed. The upshot was he agreed to permit this demonstration.

On the third day of the trial, after most of the economic and legal arguments had been made and just after the manufacturer's chief engineer had been on the stand extolling the excellent technical features of their aluminum armor design, I set up my exhibit on a card table in front of the jury box. It consisted of two one-quart mason jars, both filled with sea water taken from the exact location where the cables had been laid. In one jar there were suspended a length of copper cable tape and a length of the aluminum armor wire with intact resin covering and its bottom face sealed with wax. The second jar held the same suspended pieces in sea water except that the resin covering of the aluminum wire had several deliberately made small openings in the form of slits and holes. Again its exposed bottom face was wax-sealed. Finally, I arranged for two voltage supplies. One was from a toy transformer set at 2.0 volts, a-c; the other was from a battery also at 2.0 volts, d-c. These voltages were to be applied alternately between the electrodes in each jar.

So I described to the court the nature of this simple demonstration set and how it could duplicate the cable's failure process. The courtroom became charged with expectancy; the attorneys all moved closer and even the judge stepped down from his bench and stood with the others.

As I was applying both the transformer's a-c voltage and then the battery d-c voltage to the jar containing the aluminum armor wire with its resin covering unbroken, everyone could see that absolutely nothing was going on. Then this procedure was repeated with the second jar. Instantly the jar filled with gas bubbles, all shooting out from the slits and holes in the aluminum wire covering. After thirty seconds or so, the water in the jar became so clouded that nothing could be seen except a constant boiling in the jar and a steaming above. This was repeated regardless of whether the applied voltage was a-c or d-c, with the positive terminal going to the aluminum wire. After a little more than five minutes of this, I removed the electrodes from both jars. There was no sign of any change in the copper tape or the armor wire from the first jar. When the tape and armor wire were removed from the second jar, the tape was found to be only brightly etched in places. But the armor wire showed cover swelling at every opening. I carefully stripped off the polyethylene covering. At every spot along the wire that had an opening, the aluminum metal surface was pitted with some pits, quite deep. Of course these pits were where the metal had gone into solution partly in the sea water and partly as evolved gas.

Well, that demonstration certainly finished the case. After a brief recess during which the manufacturer's representatives and their attorneys huddled together, they returned to court and offered to negotiate a reasonable settlement. Our people, after a bit of backing and forthing, finally agreed to a proposed restitution and costs offered by the manufacturer. The trial was thus terminated to the satisfaction of both parties and the court.

I don't believe that aluminum armor wires, covered or bare, have been used on submarine cable since that time.

The Case of the Park's Power Problem

Our country has many areas of great natural beauty, some of small size but exquisite, others vast in extent and awe-inspiring. And with wisdom and foresight most of these have been set aside as parks to be preserved as they are for all times under the National Parks Service Administration and Maintenance.

This tale is about Glacier National Park, tucked away amongst the tall mountains of north central Montana and extending to the Canadian border. Between Glacier and the border is the Waterton International Peace Park, an extension of Glacier and a continuation of Glacier's beauty of snowy mountains, rapid steams, glaciers, and icy lakes. Just east of Glacier Park and running along and beyond its boundary is one of the nation's largest Indian settlements, the Blackfeet reservation.

Glacier is a fantastic one-million-acre wilderness. Those fortunate travelers who in earlier days went west from Chicago on the Northern Pacific Railroad came first to East Glacier, where they enjoyed a few days at the very comfortable East Glacier Hotel. Then they were bussed (by Charabanc) into the magnificent central park areas and stayed at the elegant Many Glaciers Hotel, with its memorable wilderness scenes. From the hotel hiking trails lead up to the mountain ramparts. The traveler would then be bussed right over these ten-thousand-foot-high ramparts through the Going-to-the-Sun Highway pass and then down along Lake MacDonald to Lake MacDonald Lodge, near the western terminus of the park. Here the aspect is quite different, with lush undergrowth and, in the summer, a profusion of wildflowers. At West Glacier, a short drive from MacDonald, the traveler again boards the train, perhaps the famed *Empire Builder*, and goes on to Spokane or Seattle. He'll not forget the natural wonders he has seen at close hand in Glacier.

Glacier was dedicated as a national park in 1910. At that time the works of man were very few. The park remained a primeval wilderness for many years. Gradually roads, maintenance buildings, visitors' centers, and hotels came into being. And with these amenities electric power, first as a spotty luxury, then as a necessity, came to the park. By the early sixties, electrical power (and communication) systems were everywhere, carrying out the almost endless tasks of lighting, pumping, refrigeration, vehicle battery charging, et cetera, et cetera, and of course operating the many appliances as they were introduced. At that time the National Park Service had decreed that all electrical transmissions and distribution systems would have to be underground—out of sight. Or if that was not possible, overhead lines would have to traverse out-of-the-way routes, with the wires having a sky-blue anodized coating.

With these great advantages to the comfort and safety of the park visitors and to the efficient operation of park administration and maintenance, there is also an ever present shadow. And that is a sudden unscheduled, unforeseen power failure. When that happens almost everything stops. And that is exactly what happened one Sunday evening in July a few years ago. The affected part of the park was in and around the Many Glaciers area, and the hotel there was jam packed at the time. All hotel grounds and area park lights were out; there was no running water in the hotel and no sewage disposal since the lift pumps had stopped. It was a most uncomfortable night for the visitors.

By early morning people from the consulting firm in Missoula who had designed the system and the electrical contractor from

Cut Bank who had installed it were on their way to Glacier. And so was I the following day, when the consulting firm called me to come out and assist the investigation. I had worked with this company on several previous occasions and had a high regard for their technical skills and integrity.

Louise and I were living at Vantage, Washington, on the Columbia River, and the day I received the call, as well as the past few days, had been real scorchers—temperatures at midday above one hundred degrees Fahrenheit. The thought of Glacier Park with its snowy mountains, et cetera, was delightful. So my wife and I packed a few things and we were off. When we arrived at Many Glaciers Hotel, we found they were not taking any guests in this emergency. The consultant firm's engineer who met us said he was staying on the Blackfeet Indian reservation in their town of Babb. That is where we engaged a room at a small Indian-run motel—spotlessly clean with very friendly management.

The problem at Glacier stemmed from a faulty cable installation. About eight months before the blackout, the Many Glaciers area had been supplied by an overhead twenty-five-thousand-volt distribution line running along the narrow valleys of the surrounding mountains and tapped into the visitors' area where the step-down transformer pads were located. In keeping with the park's efforts to preserve a primitive nature and improve the security of the electric service, it was decided to bring a twenty-five-thousand-volt underground cable into the area. This cable was to be joined to the overhead line some four miles out of the central area.

The cable selected consisted of three polyethylene insulated UD conductors, to be buried thirty-six inches deep along the shoulder of the entrance highway all the way into the pad-mounted transformer adjacent to Many Glaciers Hotel. Construction bids were let and a Cut Bank electrical contractor was selected to install and change over the service from overhead to underground. This was done and the cable had been in operation for about six months.

When I arrived at the site of the cable run, I found the contractor's work crew had determined the problem was in the new cable and were attempting to locate where it had failed. To do this they were using a "thumper" most inexpertly.

A so-called thumper is nothing more than a high voltage

direct current source that can send pulses of high voltage into the cable. Its purpose is to reduce electrical resistance at the failure location in the cable so that each pulse of the thumper can produce a very audible arc right across the failure. With the thumper operating, the electrician then patrols down the line rolling a bicycle measuring wheel along and listening for an arcing thump along the right-of-way where the cable is buried. This method of exactly locating a cable failure can work very well, but it can be abused badly and often is. If the failure is not located within an hour or so of thumping, the chances are that the voltage will be increased to its maximum and the thumping continued—sometimes overnight. This most often results in the finding of more than one failure, with all but the original one caused by the thumper. Also, the extended thumping destroys all the evidence that is useful in analyzing cause of the cable failure.

I found that the contractor at Glacier had been thumping the cable for three hours and they planned to keep it going when the crew went to lunch. I put a stop to this at once. Then, using a low voltage Wheatstone resistance bridge method, an instrument supplied by the consulting firm's engineer, we spotted the failures in all three conductors and subsequent calculations showed they all were located within twenty feet of one another at one thousand feet out from the transformer. When we measured off this distance, I was surprised to find the cable trench running along what appeared to be a reclining wall of solid rock.

We quickly had the loose shale covering of the trench removed and found the three cable conductors lying on a rocky bed only ten to twelve inches below the shale back-fill. In the fifty-odd feet that we uncovered, the cable insulations showed rock damage at many places and we counted at least seven electrical failures.

It was then that we had a long talk with the contractor and learned the worst. He said he had no trouble trenching in the cable until they reached the bank of rock near the transformer pad. To continue the cable line along the side of the road he decided to blast out the trench. And so for nearly one hundred feet along that rock shelf he drilled into the rock and placed blasting powder. The cables, which had been laid out to one side of where the trenching was taking place, were thrown up several feet with each blast. At that they managed to dig into the rock only a scant foot—and let it go at that. Then they placed the cables into this shallow, hard trench and covered them with the rock debris.

No wonder these cables showed those many deep nicks and gouges into the insulations; the wonder is that there were no electrical failures much earlier. Every axiom of proper cable laying had been violated. Normal practice calls for the cables to be buried a minimum of thirty-six inches, the cables to be laid on a bed of sand a few inches thick, the cables to be covered with a thick layer of rock-free earth before the coarser back-fill material is shoveled over the trench to bring it up to grade. As a matter of fact, these twenty-five-thousand-volt cables had been treated like garden hose.

Obviously, at least 150 feet of the three cables had to be replaced and a change in routing effected to avoid the rock bank. Fortunately, that rock surface dove steeply about fifty feet from the roadside so that it was quite easy to make a circuitous trench and bring it back to the road margin again. The question now was availability of the extra cable lengths needed. Fortunately, the park had purchased enough spare cable, anticipating extensions and emergencies, so there were three reels of it, each containing about 350 feet in their toolshed. But unfortunately, when these new lengths were spliced to the line side cables, where the damaged cables had been cut out, the other ends of the new cables were about ten feet short of joining the old load side cable ends. This meant we had to excavate the buried cables all the way out to the transformer pad, then move them back in the trench to meet the new cable lengths and splice them together.

With two crews working around the clock we accomplished the steps outlined above by close to midnight using emergency lights supplied by the service trucks. All that was left now was to move the transformer pad by some twelve feet back toward the new location of the cable ends, connect the cables into the transformer primary, and close the cut-off switches on the overhead line into which the cables were tapped, and with that, electric power would be restored.

Regardless of the very late hour, the contractor and his crews decided to go ahead and complete the job that night. However when the park senior ranger, who had been with us most of the day, learned we were going to move the transformer pad, he called a halt. He said that his orders were in effect not to permit the erection, removal, or change of location of any visible structure without the express approval of the Park Service in Washington, D.C. And none of our arguments citing that the transformer move would not cause any appearances change and that service could be restored before morning, et cetera, made any difference.

So we all left the job site with mixed feelings about the Park Service. It did seem that their bureaucratic edicts could be modified in the interest of common sense—and particularly in this emergency situation.

It wasn't until midmorning of the next day that I drove into Glacier with my wife. She went into the hotel to check on hiking around the lake, which she planned to do while I was busy at the transformer site. The consultant firm's engineer was on hand, as were the contractor and his people. All were just standing around waiting. An early call had been placed to Washington, but as of the time I arrived, there was no response to the ranger's request. About three o'clock in the afternoon we got word that Washington was considering sending someone out to the park to check on what we had done and were planning to do. I told the senior ranger that was absolutely ridiculous and asked if I could talk to headquarters. Well, finally that was arranged and after fifteen minutes of earnest conversation we got an agreement to go ahead with the move. And so by nine o'clock that evening power to Many Glaciers area was finally restored.

I should add a postscript here. You will remember when we left our home at Vantage the weather had been almost unbearably hot. It stayed that way until we were up into northern Montana, and when we drove into Glacier it was so blessedly cool that we had to put on jackets. Carrying it a bit far, when we woke up next morning at Babb there was ice on our windshield.

My wife's hike around the Many Glaciers Lake turned out to be somewhat harrowing because there was a renegade grizzly bear in the area. But that is another story for another time.

The Case of the Shocking Landlord

Avarice and stupidity, how often do these ugly traits work together to cause distress. In particular, a landlord of low-cost housing motivated by those qualities can make the lives of his tenants miserable and worse. In fact, in this instance I will relate here such a landlord, not too indirectly, caused the death of a tenant.

It was a small community in central Idaho near a large army base. Many civilian personnel and married soldiers found this community a more pleasant place to live than on the base. There was an elementary school, several churches, a movie house, and a playground with a swimming pool there. And the farmers' market every Saturday was a great attraction.

Housing was varied and fairly substantial for the permanent residents. The people from the base mostly rented space in glorified motels and the cheaply put-together apartment units. These units were small houses in groups of perhaps ten or twelve and were rented on a month-to-month basis mostly to servicemen. The owners' interests appeared to be only in making these living spaces pay a maximum amount while spending as little as possible on maintenance or repairs. In the event that repairs became urgent, be it structural, plumbing, or whatever, the landlord himself did the work, regardless of his lack of knowledge or skill at the task.

It was in such a three-room apartment unit operated by such an owner that a young serviceman and his bride of a few weeks came to live. To them, in their newly found happiness, their place was a "dream" house and its inconveniences, soiled furnishings, warped doors, and general drabness were overlooked—at first. Then as the reality of the surroundings emerged, they complained to their landlord. The front door couldn't be locked, a window was cracked, some electric outlets didn't work, et cetera. All their complaints got evasive responses from the landlord, and in fact, he did nothing about them.

From the beginning of their occupancy of the apartment just two days ago the young wife had been subjected to slight shocks when she was washing dishes at the kitchen sink. These tingling sensations seemed to come when her hand was in the running water stream or on the faucets and her body was pressed against the sink lip. This had been reported to the landlord. He told her she was just imagining things and there was nothing to do about it. But then one morning when washing dishes she received a real jolt—one that hurt. That shock made her dizzy for a few moments. After that experience her husband confronted their landlord with an ultimatum: "Get the electricity out of the sink, or we leave."

The landlord did come to the house, but he came alone and called no electrician to meet him there. I was unable to find out exactly what he did, but the serviceman's wife, who was watching him, said he clamped some heavy, bare copper wire from the gas pipeline going into the house to the pipe on the other side of the gas meter. Presumably he deduced that the shocks were due to a poor ground in the house's electric system. Since he had wired the house using the gas piping as ground (something he should not have done), he decided by jumpering around the meter the ground would be better. He was several hours at this work. When he finished he called to the wife that he had taken care of the matter and she should not have any further trouble. But he did not go in and check if what he had done had eliminated the shocks.

The wife too did not make a check. By this time she was very cautious about using that sink and thought she would wait until her husband came home. As you might guess, they found no difference. Depending on how hard they pressed against the metal rim of the sink with a finger on a faucet or in the water, they could get shocks from a little tingling to a scary jolt. They made a decision—to ask a base electrician, a good friend of the husband's, to come in on the morrow during lunch break and see what was going on. As it turned out, that was a fateful decision.

But they felt good about it and with that the husband thought they should have dinner out—at one of the two local restaurants in town. It wasn't until after eight that they returned and decided to have a bath together. He drew a fairly full tub before turning off the water and was in first, sitting with his back toward the faucets. Then his wife stepped in at the rear of the tub and to make room for her he wiggled back to the front end, right against the faucets. The instance he made contact, there was a sharp electric arc crack, a smell of burning flesh, his face turned blue, and he slid down in the tub unconscious. Somehow his wife, although a diminutive person, managed to get the unconscious man out of the tub, all the while receiving small shocks herself. Then she ran for help. Neighbors came in, some started resuscitation and others called for a doctor and an ambulance. But the serviceman never regained consciousness and died before reaching the hospital, thirty miles away. He had received a massive surge of electric current through his body.

After a time, the families of the young couple engaged a firm of attorneys and entered suit against the landlord and the manufacturers of all the fixtures in the apartment. When they filed the complaints, they still did not know the way and why the accident had occurred. What was known was that all the metallic water drains in the house were at full line voltage, 120 volts, to ground and that all the water service piping and faucets were grounded. This was determined by a power company electrician who then shut off all power to the house.

That was the situation when I was asked by the families' attorneys to investigate what caused the electric shocks. I spent half a day discussing the background of the accident with our attorneys. They knew the details of the factual happenings and a great deal about the landlord and how he operated his rental properties. What I could not understand was the power company's report about the drain lines being energized. Either the electrician who checked the house was mistaken or there were some very bizarre tie-ups involved.

Arrangements were made for me to visit the still closed house, and I requested the power be turned back on. Of course the first thing I did was put a voltmeter across the sink and tub faucets and drains and then I checked for grounds. Sure enough, the faucets and water piping were well grounded, but the drains everywhere in kitchen and bathroom were at full service voltage. How could this come about?

There was a crawl space under the house, and I got into it to see where the drains went. There I found that the metal drainpipes from the kitchen, bathroom basin, tub, and toilet all terminated in a large plastic pipe, which continued underground to the area sewer lines. That meant that the drain system had no definite ground, which was not too unusual. It also meant that if somehow the drain lines were energized, they would remain so without blowing a house fuse or opening a branch circuit breaker. But again I asked, *How is it possible for these drains to be energized—and energized at full voltage?* Not knowing what else to do, I pulled all the appliance plugs out of their receptacles one by one and checked for any change in the drain voltage. There was no change. Nevertheless, somewhere in the house a drainpipe was in metallic contact with a line power wire. This had not occurred outside the house, since when the main switch to the house was opened the drain voltage disappeared.

Then late in the afternoon we received a break, which eventually solved the puzzle. A recent tenant who had occupied this house a few days just before the soldier couple, seeing lights within, stopped by. He told us much about the landlord and his slipshod maintenance, but what was most interesting was the tenant's reason for leaving the place. On his first day of residence—it was a cold fall day—he found that the apartment gas heater was in such a poor condition that they could not get the place warm. After a great deal of complaining, the landlord finally brought in another heater, not a new one but one taken out of a vacant unit. Of course he did the removing and installing himself. The result was a poorer performance than before. The flame safety switch would repeatedly kick off the burner after only a few minutes of heating. The landlord promised to come back and fix that, but after waiting two days with nothing done the tenants left.

This information immediately turned my attention to the heater, which I had overlooked before. It was a conventional wall-type gas heater such as used in motel bedrooms. It was located in the living room and it had a metal exhaust stack going up through the ceiling and out to the roof. The heater appeared to be designed for use with either natural or bottled gas and incorporated a blower fan motor. A flexible electric cord extended from the base of the unit and was plugged into a wall receptacle. The receptacle was a modern type with two line and a ground socket for a standard three-blade plug. However, the plug on the heater cord was a simple two-blade nonpolarized plug; evidently a hardware store item replacing the original molded-on plug.

As I was examining the heater I inadvertently put my hand on the metal switchbox containing the blower switch and immediately felt a tingling shock. The toggle switch itself had an insulating handle so there was no shock felt when turning the blower on or off. But the metal frame of the heater *and its stack* were continuously energized, and my voltmeter showed 120 volts to a waterpipe ground. Obviously, the hot wire to the motor had a bare spot somewhere and that spot was in good contact with some part of the heater frame. Also, the frame was ungrounded because the power cord plug had no grounding blade. If there had been a proper three blade and the ground wire in the cord were connected properly to the heater frame, the inadvertent bare spot contact would have immediately blown a branch circuit fuse or breaker. But that didn't happen and it couldn't happen under the circumstances.

Now we had an energized heater frame and exhaust pipe, but how did that get in to the house drains? I soon found the answer. The exhaust pipe terminated on the roof, which was covered with sheet metal roofing and at the time some snow. The metal flashings around the pipe put it in good contact with the roof, and so they too were energized at approximately full voltage.

The mystery of the drains was thus solved. In that tub on the fateful night, the husband was sitting deeply in water that was energized by the drain fitting. It was an enameled tub so that as long as he didn't come into contact with any grounded object he would hardly feel any shock tingle. But the faucets were well grounded so that when he backed into them he was electrocuted. Normally, shocking currents are limited by the relatively high contact resistance of the skin and/or the clothing. The internal body pathways have relatively low resistance. So if a body is bare and wet or immersed in water and under these circumstances is subjected to a reasonably high voltage, the current flow through the body will be very high, since the contact resistance can be vanishingly small.

We called in a licensed electrician to take the heater wiring box apart. There in the close confines of the metal one of the wires going to the blower motor switch had been stripped of its insulation for several inches, presumably to make the connection easier, and part of the bare wire was lodged firmly against the box wall. And as the plug sat in its receptacle, that stripped wire was the hot one.

The case never came to trial. The landlord's attorney strongly advised him to negotiate a settlement. He foresaw the possibility of some criminal elements emerging against his client in a court trial. The families eventually did agree to a settlement, and there the matter ended as far as I know.

What a sad affair and what a pity, a young man shamelessly sacrificed to greed and stupidity.

The Case of the Mysterious Mines

In 1943 the war had come to our coasts. Isolated instances of the U.S. ship sinkings at sea by German U-boats had been reported from time to time. These were mostly cargo vessels carrying supplies and munitions to the European fighting fronts. As escorting by naval vessels improved in experience and numbers, the at-sea losses decreased drastically. But then almost overnight, we started to lose oil tankers plying our coastal waters from Philadelphia up to Boston harbor. In none of these cases were any submarines involved, so that the only reasonable conclusion was that mines were responsible. These were obviously laid by German submarines in the waters of our busy coastwise shipping lanes and were proving to be extremely effective. At one time during this period we were losing a tanker every day.

Of course what minesweeper vessels and gear the navy had at that time were put into operation, notably around the ports of New York and Boston. These were conventional M.S. vessels using conventional mechanical sweeping gear to gently pick up and retrieve the usual floating or shallow tethered mines for later disposal. The sweepers were also prepared to use gunfire to detonate these mines as circumstance permitted.

But no mines were found, even after painstaking coordinate search procedures with assistance from aircraft. This was a very perplexing situation. How then could our ships be blown up when there were no mines to be struck and detonated?

Then we learned that Great Britain was also having serious shipping losses in and around its harbors, obviously due to mine action. And just as we had experienced, their minesweepers could find no mines. The mystery was proving to be a very costly one; ships were continuing to be sunk by the same inexplicable agency. What deviltry had the Germans concocted now?

Two events, one off our coast and the other on an isolated English beach, solved the mystery. The first of these was the blowing up of a U.S. minesweeper off of Boston harbor. This minesweeper was a large double-purpose ship able to destroy submarines as well as sweep for mines. It was equipped with very sensitive sound propagating and receiving gear (asdic) to detect location and distance to submerged submarines. Just before the ship was sunk, the asdic operator had picked up echoes from a small object submerged about seventy-five feet. The vessel was moving so slowly that the asdic operator could see the object range decreasing slowly for a moment before the explosion took place. Fortunately, most of the minesweeper crew were rescued and in particular the asdic operator was uninjured and able to report his echo pickup.

In England a large, steel, ball-shaped object had washed up on one of the barbed-wire beaches. As it was suspected that it was some type of mine, demolition experts were called to deactivate it. They had never seen anything like it, and as the object was slowly disassembled their wonder grew as the complexity of the electronic and magnetic detonating system, for half a ton of explosives was revealed.

When the combined U.S. and British naval intelligence units had completed their study, it became apparent that this German mine required no direct contact with a ship for its detonation. It could be tethered as much as one hundred feet below the sea surface and still be effective. The initiation of the detonating mechanism was by a delicate magnetic device that when the mine was placed in position was balanced against the earth's magnetic field at that location. Passage of the steel hull of a ship would appreciably change the earth's magnetic field at the mine and thus unbalance its magnetic device. This then set in motion the firing mechanism. It was also apparent that the magnetic unbalance had to proceed slowly from low to maximum, approximating a ship's passage, before it armed the detonating train.

And so we learned about the new magnetic mines and immediately set about finding how to neutralize them. Two schemes taken together finally did remove the terrible threat of magnetic mine sinkings. Obviously these mines could not be retrieved nor could they be detonated by gunfire from a safe distance. What we came to after much technical theorizing and experiment was to:

1. Equip our ships with a girdle of cables through which a direct current of some magnitude was circulated. The direction of current flow and its magnitude was adjusted to just balance the ship's inherent magnetic field—in other words, to neutralize any external magnetism the ship might normally have. To make this adjustment precise, at several of our harbors sensitive magnetometers were imbedded on the bottom, which could provide a remote record of the number of gausses of magnetic field intensity a ship had as it moved on a prescribed course overhead. Then adjustment of the girdle current on the ship was made until the magnetometers registered zero. The process was known as a degaussing operation. A properly degaussed ship could slide by a magnetic mine implant harmlessly.
2. The second scheme, also ultimately successful, was to detonate a mine by a surface sweeper sufficiently distant to be unaffected by the explosion. I was intimately involved in its design, development, and field testing.

At that time I was a research engineer for a large wire and cable manufacturer. We were heavily involved in turning out enormous quantities of insulated cables for the navy's ship degaussing program. In charge of the program for the navy was the then Captain Hyman Rickover, and during the course of this degaussing program I had frequent contacts with the captain and his staff.

Experimental studies by the navy had shown that a magnetic mine could be set off if a heavy, on-off pulsing electrical current flowed in its vicinity. To carry this out practically and with impunity it was conceived that a minesweeper would have a large d.c. generator with several thousand ampere capability and suitable on-off timed switching gear. The generator output would feed into a long, well-insulated, trailing buoyant cable at the end of which was a shorter bare conductor electrode discharging current into the sea.

My company was given the problem of designing and manufacturing the buoyant cable and its electrode, and that was the task I was directed to in all haste. We looked at a number of possibilities, the simplest of which was to have a large-diameter watertight hose on the outside of which the necessary number and size of copper strands were laid helically and over that thick rubber insulation and a strong jacket of asphalted braided jute. This simplest design was the one adopted. The sealed end of the buoyant cable was fitted with a connector to which a flexible bare stranded cable could be attached to act as the sea electrode.

With navy approval we started manufacture of the cables and samples from the first few runs were fitted up for field trials. For these tests, 1,500 feet of buoyant cable plus 150 feet of bare electrode cable were wound on a huge reel mounted at the aft end of the sweeper vessel. As the ship got under way, the cable was unreeled and entered the water by a long, trailing line. With the inboard end of the buoyant cable connected to the generator and the pulsing current started, any mine detonations would then occur at least a safe fifteen hundred feet astern of the vessel.

The tests were very successful and demonstrated that we could rid our shipping lanes of magnetic mines. However, there was one problem. When a mine was detonated, like as not the electrode section would be torn off the buoyant cable, destroying the latter's end seal. When this occurred water would immediately enter the cable's hollow core and it would sink. Under that condition the cable could not be retrieved and had to be abandoned—all fifteen hundred feet of it. Also it was found that small craft inadvertently crossing over the long trailing cable damaged it sufficiently to open its protective covering and again the cable filled with water and sank.

My solution was simple but tricky to incorporate in the cable manufacturing process. We bulkheaded the cable. By extruding vertical sealing barriers every ten feet in its hollow core, the cable could sustain a number of severe damages and still remain afloat. We had to set up special X-ray equipment in the manufacturing line to visually check the barrier closures and make sure the bulkheading was effective. As we gained experience and developed improved extrusion techniques, making the buoyant cable became routine and we supplied the navy with many hundreds of magnetic minesweeping cable units.

With these new trailing cable devices the navy went to all-wooden-hull minesweeper vessels, designated YMSs. By this precaution a YMS had of itself little effect on altering the earth's magnetic field and so had no need for a degaussing girdle. Another large improvement in sweeping efficiency came when two YMSs moving parallel to one another and separated by about twenty-five-hundred feet pulsed their sweeping currents alternately. For the pulse period when sweeper no. 1 was sending current into its electrode, sweeper no. 2 current was off. At the next pulse, sweeper no. 2 supplied the electrode current in the opposite direction while sweeper no. 1 current was off. This improved operating system did much to clear all of our coastal shipping lanes of magnetic mines for the rest of the war.

Some of my experiences during the work on magnetic minesweeping were grim. For a day of sea trials I reported to a naval pier at the foot of Broadway in New York a little before 6:00 A.M. There the YMS was all ready to take off, just awaiting some navy brass and a few engineering experts. By 7:00 A.M. we were well down in the harbor and starting through the Narrows. When we met the ocean rollers, the little vessel pitched and yawed sickeningly, like a bronco, but as soon as the long, buoyant cable was unreeled into the water, it had a marvelous stabilizing effect. Then as we got under way along the prescribed sweeping course, we might be so unfortunate as to pass through the ugly, sickening debris of one of our blown-up ships. There was always a great oil slick, and in it were all manner of floating pieces and garbage from the destroyed ship. We would slowly move around the area looking for bodies or possible survivors.

On the lighter side, I recall several occasions when Captain

Rickover went out with us on these trial runs to check operation and management of the long, buoyant cable. As we took off and stood down the harbor, he was very much in evidence checking generator output and how we were preparing the cable for reel-out, giving instructions and orders to us and the ship's officers, et cetera. But as soon as the ocean swells started to lift and drop the vessel, Captain Rickover disappeared; he was not in evidence for much of the trip. I later learned he was subject to seasickness and preferred to wait out the roughness in his stateroom. There were rough sea times when I too would have preferred to be lying in bed, but no such luck for me.

I do want to say here that my admiration and esteem for Captain Rickover was very great. During the traumatic time of the magnetic mine sinkings he organized the various teams that worked to eliminate the menace. It was his leadership and hard driving and never taking "no" or "cannot do" for an answer that were largely responsible for our success.

The Case of the Hotel Air Conditioners

The rather mysterious events I will relate here took place at a popular resort hotel in a south-central Washington city bordering the Columbia River. Its proximity to this great river made the city a mecca for beach and boating enthusiasts, and during the spring, summer, and fall its tourist facilities were usually taxed to the limit. In particular, the special events such as regattas, speedboat races, water-skiing exhibitions, and the like filled the hotels, motels, and restaurants. The resort hotel of this story was the largest and most luxurious in town, and its clientele were generally well endowed with wordly goods.

On a hot July day a few years ago, a national speedboat racing competition had just been opened and the trial runs were under way. The resort hotel was completely filled with boat owners and their guests. Just before noon, the lights in the lobby and offices started to flicker strongly and the telephone switchboard went dead for outside calls. This continued for about twenty minutes. When the situation returned to normal, calls were made to the local municipal power department and the telephone company's office reporting the disturbance.

A driver in a utility trouble truck came by shortly afterward, looked around briefly, found nothing he could see wrong, and departed. The electric system operated without further incident until about 3:00 P.M., when suddenly there was an explosive sound and about one-half of the electric lighting went off and the telephone service quit entirely. It was a partial blackout! One can imagine the concern of the hotel manager, faced with a full house of 350 demanding people, in a predicament of this kind. His frantic calls (made outside) to the power utility this time brought several truckloads of service men, a construction foreman, and the department's distribution engineer.

This hotel was supplied by a quite large transformer bank consisting of three transformers connected to supply three-phase power with its incoming primary at fifteen thousand volts. The transformers' secondary supplying the hotel loads was also three-phase, with the lighting systems and electric heaters taken off at 120/240 volts single-phase and large motor loads at 415 volts, nominal three-phase.

It was close to 6:00 P.M. when the utility men located the cause of the power failure in the transformer vault along the side of the main building. Here they found one of the transformers had failed and become short-circuited. This had blown the incoming primary line fuse on that transformer's phase, but not the fuses protecting the other two transformers.

The vault had a relatively small iron grill cover that could be removed, and this is where the service men got down to examine the transformers. Getting the failed unit, which was large and heavy, out of that vault was a difficult job because the cement walk had been paved over to the edges of the grill. Thus some extensive jackhammering had to be done to break out a large enough area to pass up the transformer. The second major problem that night was to find another transformer to replace the badly damaged one.

All this was going on under the continuous importuning of the hotel manager for haste and the caustic comments of many of the hotel guests who formed a standing audience. The construction foreman made a kind of offhand comment to the manager that it would be well if they shut off all power to the hotel until a replacement transformer could be connected in. At this the manager literally hit the ceiling. He shook his finger in the foreman's face and shouted, "No, no, no!" And so the supply of power with only two phases active continued.

At about 10:00 P.M. the damaged transformer was finally pulled out of the vault, and just at midnight a replacement unit was trucked into the hotel yard. It was quite obvious even to the remaining onlookers that the "new" transformer looked different from the damaged unit in size, shape, and bushing complement. Nevertheless, they lowered it into place and by 2:00 A.M. full power was restored to the hotel.

In the meantime, while these power transformer maneuvers were going on two telephone repairmen had been working on the telephone switchboard. They told the manager that the telephone line surge protector had blown out and this had caused short circuit of a bridge rectifier—all of which could be readily replaced next morning from their local parts inventory.

I should also relate what was going on in the hotel during that afternoon and evening. Recall that the lights had remained on in many of the rooms and corridors and this included the kitchens. (They were off only for a half hour at a time while the failed transformer was disconnected and while its replacement was reconnected.) However, none of the room air conditioners would work, although they had been in operation continuously prior to that afternoon's trouble. Even though it was a very hot day and room temperatures had risen into the eighties, those guests who were in their rooms did not complain because it was assumed that once the hotel's power was restored the air conditioners and everything else would be working again.

You will recall that at the time the damaged transformer was discovered, the utility construction foreman had, in an offhand manner, suggested that they shut off all power to the hotel and this the manager refused. However, that remark was heard by the hotel's all-purpose handyman and, being somewhat knowledgeable in electrical equipment, he interpreted that to mean that any three-phase motors could be overheated and damaged when forced to operate on only two phases. As a result, he and his helper went all over the hotel switching off all three-phase motors and these included *all* of the hotel's guest rooms and assembly areas' air conditioners. It took them several hours to do that, so from the time the blackout occurred many of the air conditioners had been switched on for perhaps five hours before all could be shut off.

Restoration of electric service brought no rest or peace to the weary manager. Now in the early hours of the morning calls were flooding the switchboard with guest complaints that their room air-conditioning units were not working. A hurried check by the manager and the night clerk confirmed that the conference and restaurant air conditioners as well as those in the guest rooms could not be operated.

With daylight and most of the hotel's normal chores begun, a detailed count showed 142 air conditioners out of service and presumed damaged in some manner by the loss-of-power events of the previous day. An appliance repairman from a reputable large local dealer spent several hours examining a number of them. He told the manager that in each unit he examined the compressor motor insulation was burned "to crisp" and winding turns melted apart in places. He thought that although it might be possible to repair some of the units, on the whole it would be better to install new compressor motors. However, the dealer pointed out that buying new compressor motors, disassembling the damaged motors from the units, and installing the new motors would cost more and take longer than simply ordering entirely new air conditioners of the same model and size.

With such a program the hotel was facing a very substantial outlay—to the tune of seventy thousand dollars. Another blow came when the manager called the hotel's insurers and learned that there was a clause in the policy relieving the insurance company of damage to the electrical system.

Faced with most of the hot summer season still to come and their solid bookings for months ahead, the hotel decided to repair what was repairable and purchase new units for the rest. Inquiring at the city power department brought the response that they were in no way responsible for the loss of the air conditioners and that it was the hotel's mishandling of its own equipment at the time of the transformer failure that caused the motor burnouts. With that ultimatum from the city, the lodging chain that owned the hotel decided to bring suit against the city, and to prosecute it they called in a local law firm, which I will designate as ABC.

The legally filed complaint was to the effect that a major malfunction of the city's electric power distribution equipment supplying the hotel was (had to be) in some way responsible for loss of all the hotel's air-conditioning units. The city immediately brought in a motor expert consultant to assist in their defense, and ABC called on me to assist their client.

A great advantage in my investigation in this case was the fact that its events had occurred only a few weeks before. The various people who were involved were still around to be questioned, and their responses were from fresh memory. Also, some of the failed air conditioners were still available for examination, and in particular, the utility had not yet junked the damaged distribution transformer.

I spent several days in the city, talking to and interviewing the hotel manager and staff as well as the utility service men and their distribution engineer, and he provided a schematic print of the hotel's power supply system. Also, with the motor repairman originally called in by the manager to look at the air conditioners I took apart one of the failed units and examined it in detail. My request for a print of the hotel interior wiring came to nothing, as none could be found. The hotel was really quite old and had been remodeled completely at least on two occasions, and their files had not kept up with the changes.

I had better luck with my request to examine the utility's failed distribution transformer. At first we were told that it had been junked and was unavailable. However, with some persuasion from the attorneys, the utility decided the transformer could be viewed in their equipment yard on the outskirts of town. So I was able to examine it and make some resistance measurements on its windings and terminals—all of which were quite revealing.

Eventually some interrogatories from both parties were submitted and answered. Eventually a number of depositions by involved personnel of both parties were taken. And eventually depositions by the utility's expert consultant and by me, the hotel's consultant, were taken.

From the above documentation the utility's position appeared very clear. Briefly they contended:

1. That the failure of their transformer was an "act of God," for which the utility had no responsibility.

2. That it is well known that three-phase motors forced to operate on reduced phases will overheat and "burn" up in a matter of a few hours.
3. That although warned by utility personnel at the time of the transformer failure, the hotel persisted in maintaining the two-phase supply to its interior wiring for many hours.
4. Since due to the hot weather most of the air conditioners in the hotel were left on continuously, they were overheated and damaged by operating for a long period on the two-phase supply.

On the face of it, the utility's analysis of the air-conditioner failures appeared very solid and indeed it was what I had in mind when I first heard of the case. A motor, any kind of motor, during the fraction of a second when it is switched on draws from ten to fifteen times the amount of current than when running up to rated speed. At or near its rated operating speed, a motor generates a counter-voltage that opposes the applied voltage, thus reducing the current drawn to the low value consistent with its design. If for some reason that input line voltage is reduced, the motor will either turn over sluggishly or even stop entirely and thus draw high current well beyond design so that overheating is inevitable. That is what can happen when a three-phase motor has lost one or two phases; it will overheat and destroy itself if left on.

Was the utility right in its contention as to what had happened? No, it was not—by at least 90 percent. Here were the results of my investigation, which pointed to another cause to fit the evidence I found. Briefly:

1. Four hotel guests who happened to be in their rooms at the time of the mishap told me they heard a loud "pop" at that time and then smelled insulation burning.
2. A number of the air-conditioner room switches were found arced over. This occurred in rooms where the air conditioners were supposed to have been off all afternoon yet were found to be damaged and inoperative.
3. In the main electrical panel board of the hotel a number of the branch circuit breakers were found to be arced over.

4. The molten drops of copper along the windings of the damaged air-conditioner compressor motors could not have been caused by overheating but could by high voltage arcing.
5. The telephone system surge arrester was found to have arced-over during the outage.
6. The city owned and was responsible for the whole of their service entry to the hotel, including both the high voltage primary *and* the low voltage secondary utilized in the hotel. The city had *not* installed secondary surge arresters at the entrance to the hotel's power panel.
7. My tests on the failed transformer conclusively showed that its primary (high voltage) winding was solidly short-circuited to its secondary (low voltage) winding.
8. The failed transformer was over twenty years old and had been moved about a great deal in various service locations.

There could be only one answer to the circumstances given above. And that is the transformer failed because the insulation between primary and secondary windings broke down. This sent a surge of high voltage, perhaps peaked at as much as 25,000 volts, into the hotel wiring and this was instantly followed by the full primary (to ground) voltage of 8,660 volts, sustained until that phase's fuse melted out—a matter of five seconds or more. This is what caused the arcing over the switches and arcing in the compressor motor windings, nominally rated for six hundred volts maximum. Thus all the damage sustained by the air conditioners was done in the first five seconds following the transformer short circuit. There was no overheating due to phase deficiency as the city contended, because the motor windings were already open circuited by the high voltage arcing.

As the case entered its final phases it began to appear that the city might be willing to modify its hard no-responsibility stand. About a week before the date set for the trial, the opposing attorneys met together to seek a possible settlement. The compromise they reached was presented separately to each of the litigants. My client, the hotel chain, asked me to sit in on these negotiations, and I did. Starting from a 65 percent responsibility agreed to by the city in the first round of negotiation, we finally ended with an agreed city responsibility for 80 percent of the damages.

And so the case was settled, if not to the complete satisfaction of the hotel and the city, certainly to the satisfaction of the opposing attorneys.

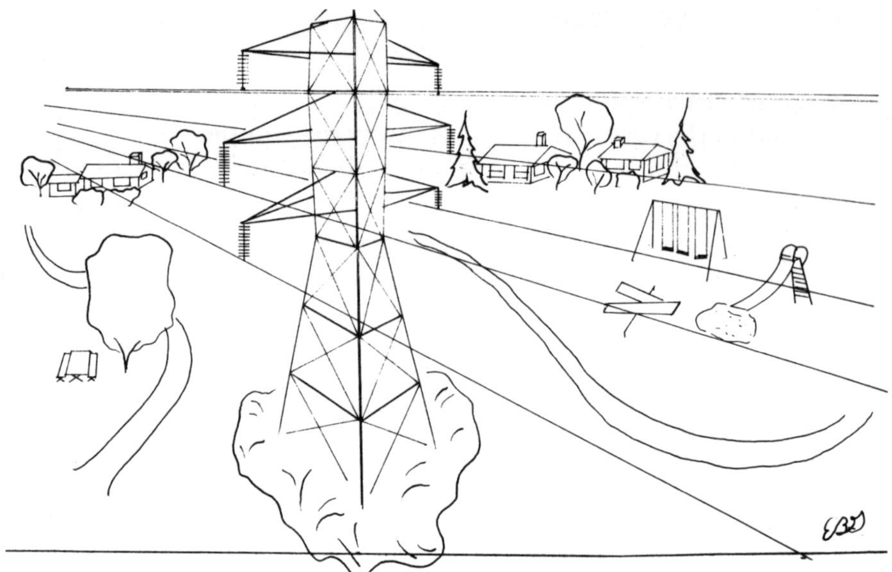

The Case of the Forbidden Climb

It happened about ten years ago on the west coast in a large city with many parks and recreation areas. A number of the parks were actually small neighborhood affairs where local children and families could picnic. One such neighborhood on the east side of town was a particularly popular place for the children of the rather well-to-do community surrounding it.

When this park was created the city had to provide an easement through it for an existing overhead 138,000-volt transmission line owned by the local power company. This line had existed many years before the park came into being and, in fact, was built at a time when that section of the city was quite rural. As streets and residences spread out farther to the east, they engulfed the power line and this caused some unsightly easements, some of considerable length, through the neighborhoods it traversed. Removal of this power line was in the plans of the power company, but at the time I became involved it was still there.

Visualize the high voltage transmission line running through the little park supported on three steel towers, each approximately sixty feet high. The three crossarms on each tower held two three-phase circuits, one circuit on each side of the tower, each consisting of three energized conductors vertically spaced about five feet apart. The voltage between the spaced conductors was 138,000 volts and between any conductor and the grounded steel 79,674 volts. This then was a typical three-phase grounded neutral system.

The unfortunate occasion that brought this park and its high voltage overhead line to my attention and concern was the near fatality of a young boy who climbed up one of the steel towers. He was fourteen years old, and he and a younger companion had apparently climbed up to the top of a favorite tower several times before. For an energetic youngster it was no great feat to make such a climb. In the usual design, the steel tower legs are braced by numerous diagonally placed lacings whose spacing decreases as the top is approached. These lacings can act as rough ladder steps, and with dexterity and good handholds the tower can be climbed. Of course when a lineman climbs a steel tower he uses a light twelve-foot ladder, which gets him up to built-in tower foot spurs that go up to the top.

As the story of the accident unfolded, it appears that Tom and his companion had always climbed up and then down on the inside of the tower and while at the top had sat there smoking pot and enjoying the forbidden adventure. I would like to emphasize the fact that their prior climbs took place inside the tower's encircling, grounded steel structure. In effect they were inside what is usually termed a Faraday cage, in which the electric fields of the energized conductors hanging on the outside of the tower could not penetrate but were essentially terminated outside on the steel.

On the day of the accident the two boys had again climbed up on the inside of the tower and spent a while at the top and then Tom decided that they should climb down on the outside of the tower, since it was different and, he thought, easier going. His companion was against doing this and started down the usual way. But brave Tom climbed over the top bars of the tower and started to move down the outside steel lacings. Just a few feet down brought him opposite and between the first energized conductor and the tower and in the strong electrical field created there. At the first shock that passed through his body he lost his hold and fell directly down. As he passed each of the vertically spaced conductors an electric flashover occurred between conductor and body and between body and tower steel. He finally fell into a heavy growth of blackberry bushes, and that saved his life.

The rescue squad arrived in record time. They found the boy badly burned, with both legs and an arm broken and a fibrillating heart. They managed to bring the heart out of fibrillation and got him to the burn center hospital fast. As it turned out, Tom, although alive, is permanently confined to a wheelchair and carries the massive disfiguring scars of arc burns.

A month or so after the accident, Tom's parents instituted legal action against the power company. Major basis for the suit was the claim that the high voltage power line traversed a well-used public park and represented an accessible, dangerous, "attractive nuisance" to the children who habitually played there.

As might be suspected, the power company rejected these claims and prepared to contest the suit. They hired an eminent law firm that in turn asked me and one other engineer to look into the technical aspects of the accident.

As the case developed we learned that the plaintiff was also prepared to claim that the utility should have been able to de-energize the transmission line very quickly—at the first arc discharge—thus reducing the burn injuries to a great extent. Also, it would be claimed that the spacing of the conductors on the towers was shorter than standards required, et cetera, and that this was again related to the severity of the injuries. My colleague confined his studies to the power flow operation of the line and the interrupting capability at its controlling substation. My own inquiry was directed to the tower itself and why and how the flashovers occurred during the boy's fall.

A major question was whether a normal fourteen-year-old boy had any appreciation of the danger in coming into contact or just close to high voltage electricity. Tom had explained to his parents that he thought that if he didn't actually touch the line he couldn't be hurt as he climbed down the tower. Hadn't he climbed up and down inside the tower a number of times before with nothing at all happening? Also he had had similar "touch or not touch" experiences with the electric outlets in his house.

In my deposition I explained to the plaintiff's attorneys how in the strong electric field existing between the energized conductor at high voltage and the grounded tower steel the air spacing was set wide enough so that no flashover would take place under most conditions. However, if a body, of any kind is placed in that space between conductor and tower the voltage across the now most reduced air space will invariably become higher than the breakdown strength of the air there and a flashover will take place—followed almost instantaneously by a flashover between the body and tower across the other space. So it is not necessary to touch the line conductor for a discharge arc to take place to a body passing between it and the tower.

Further investigation disclosed that the clearance or opening time of the circuit breakers was on the order of one tenth of a second, which was also about the same time required for the falling boy to pass through the electric fields of the three conductors. Also, the tower design as to crossarm and conductor spacing and bracing features was in accordance with the standards in effect in the 1940s when this line was built.

We did find that none of the towers in the park carried a "High Voltage Danger" or similar readable sign. In checking into this, it appeared that such warning signs had been posted on each tower but were removed (stolen) from time to time as souvenirs. Replacement was made at only infrequent intervals as and when maintenance linemen happened to be working in the park.

Putting all the positive and negative factors together and taking heed of its consultants' recommendation, the power company decided to negotiate a settlement of the suit rather than go to a trial. This was done and the case was closed.

During the course of my investigation, my son and I built a one-to-two scale model of the tower using Erector and Mechano set parts. The tower was complete with tiny insulators holding short lengths of stiff wire to simulate the transmission conductors. Also, we had a small figure almost to the same scale to represent the boy. The model was used to show how the tower's design features were related to the electrical load it carried. We were also going to use the model during the court trial. However, we found a better use for it. I asked Tom's family if, when he was able, it would be possible for him to take the model around to various boy Scout meetings, Boys' Clubs, public schools, and the like and speak on the dangers of playing around electric power lines of any kind. Parents and Tom agreed at once, and as far as I know, this was carried out for several years.

One more item to add: several months after the accident the power company rebuilt the park towers with wider crossarms and conductors and also put a stout barbed-wire fence around each tower.

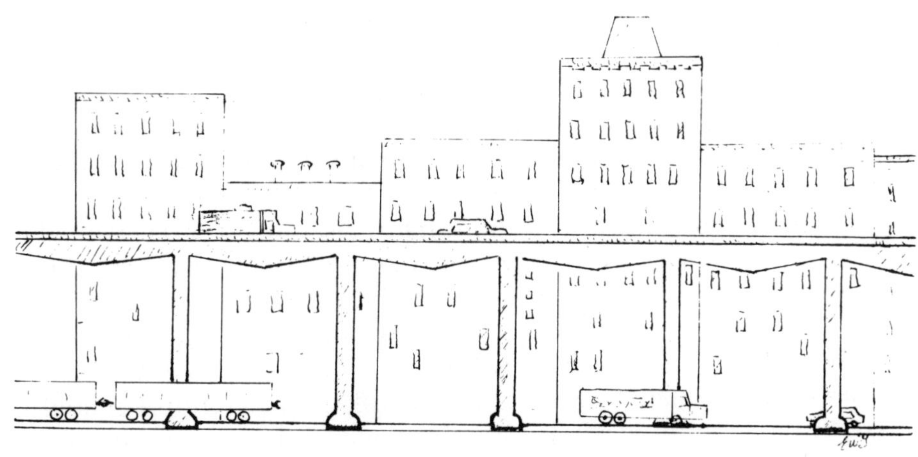

The Case of the Forgetful Supervisors

One night back in 1977, an oil company's huge tractor-trailer rig, loaded with sixty-five thousand gallons of aviation fuel, was traveling south on an elevated roadway skirting the waterfront in a large west coast city. It was heading for the airport. The roadway was wet—but not enough to reduce the oily traffic film.

Just about where the elevated passed alongside the downtown business district, for some unaccountable reason (at that time) the trailer skidded sideways, then rolled over on its side, smashing the tank walls and sending thousands of gallons of fuel cascading down on the street below.

Moments later the running fuel caught fire and started a holocaust of flame raging all along the lower street between it and the overhead structure. The flames rapidly spread to the buildings lining the east side of the street, and parked autos and railroad cars on a siding at the street level were also engulfed in flame. Many buildings were soon ablaze, and even the steel structure of the elevated roadway started to warp and the concrete to blister severely.

But that wasn't all. Attached to the underside steel of the elevated and running along its length were six very high voltage cables. These were part of two three-phase 230,000 volt circuits that supplied electric power to most of the downtown area, including the waterfront facilities and all of the marine fire fighting stations.

Since it became my responsibility, as requested by the oil company's attorneys who called me into the case, to deal with this aspect of the fire, I should describe these power cables in more detail. They were of a design known as oil-filled, a type of cable commonly used for such high voltages. The copper conductor is hollow and insulated with many layers of special oil-impregnated paper tapes, and the whole assembly is covered with a continuous extruded lead sheath and other protective outside coverings. Thin insulating oil under positive pressure from pressure-producing reservoirs at one or both ends of the cable run is contained within the hollow core of the conductor.

The purpose and the necessity, of this design is to maintain the cable insulation under a positive oil pressure at all times. Gas or vacuous voids in the insulation at this high operating voltage would spell disaster. As the cable heats up under load, the oil expands and is expelled into the reservoir. When load is dropped and the cable cools, the oil contracts and to compensate for this, oil flows into the cable from the reservoirs. Thus despite load changes, the cable insulation is always under oil pressure. To repeat, at no time can an oil-filled cable operate properly if the oil pressure is lost or the oil becomes contaminated with moisture as, under the worst condition, when atmospheric air is allowed to enter into the cable structure.

The six oil-filled cables hanging under the elevated roadway all ruptured in the fire and completely burned out over a distance of more than two hundred yards. This plunged most of the downtown area in darkness. And from the ruptured ends of the cables solid spouts of insulating oil under pressure added their high calorific fuel to the inferno. Thus the cable oil reservoirs from both remote substations, each of two to three thousand gallons, became part of the combusting fuel source and were an important factor in sustaining the conflagration and adding to the damage.

Unfortunately, at the time of the fire the municipal electric department's line crews and electricians were on strike. So it became necessary to call out all of the electrical supervisory staff to switch over the damaged cables and temporarily restore power to the city from other circuits.

The fire was finally brought under control after some eight hours during which every piece of fire equipment in the city was called in.

While it lasted, all of the oil in the cable reservoirs had run out, and as a result, the entire mile and a half of oil-filled cables was rendered unusable due to loss of insulation oil and atmospheric contamination. Most of the buildings on the lower street's east side were badly damaged for two city blocks. Some eighteen parked cars were totally destroyed, in addition to eight railroad boxcars. Also, the street's macadam pavements were melted into grooves and hollows and there was considerable damage to the understructure of the elevated roadway.

The fire and its aftermath of damages triggered a large number of law suits by many plaintiffs. Smaller suits involved burned out automobiles and the railroad's damaged freight cars. Others involved the manufacturer of the tank truck and the manufacturer of its specially designed wet-weather braking system. Also, a number of occupants of the buildings involved in the fire brought suit. However, by far the largest legal action against the oil company was entered by the city for an amount to cover the restoration of the damaged buildings (which it owned), the damages to the street surface and elevated structure, the cost of an installed, new, complete run of oil-filled cable, and the cost of an immediate new, temporary four-mile pole-supported line that they placed along the overhead roadway to carry the power needed until the much slower delivery cable replacements could be effected. All in all they were asking for over $6 million in damages.

It was at that time that the oil company asked me to look into several aspects of the city's suit, principally to determine if the city had exercised good technical and economic judgment in assessing the high voltage cable damage and if its decision to build a very expensive temporary overhead pole line to provide the downtown service, as it had, was necessary. Why during the fire had the oil-filled cable oil reservoirs been allowed to run out completely? That, particularly, was a question that bothered me. In a mile and a half of oil-filled cable there are usually three or four joints, since manufactured continuous lengths for this large size cable couldn't be over two thousand feet or so. There are two types of such joints. A through-joint permits the oil flow from the reservoirs to pass unrestricted through the joint. A stop-joint incorporates the means to readily seal off the oil flow at that position if desired. So I started to look for the joints in the cable run, where they were, and of what type.

What I found was most interesting. Checking the as-built plan of the original cable installation showed that each of the six cables had three joints in its length and that all joints were stop-joints. Furthermore, in each cable there was a stop-joint located only a few hundred feet to either side of the point where the fire started.

In other words, there were built-in stop-joints a short distance to each side of the fire where the cable oil flow could have been shut off, thereby greatly reducing the fire extent and also saving most of the cables. The city supervisory employees evidently forgot all about the stop-joints and so allowed the cable oil reservoirs to drain completely into the fire area and leave the rest of the cables to be contaminated.

Also, it turned out that had the oil flow been properly shut off at the stop-joints, there would have been no need for that temporary pole line hastily built by the city. The city had in storage sufficient spare oil-filled cable lengths that could have been spliced in fairly quickly to restore much of the needed downtown power.

The brief that I prepared for my client covered the above findings in detail. I also took occasion to point out that the hanging of such an important cable system in unprotected messenger loops under the elevated bridgework exposed it to all manner of hazards from the street below.

Finally, as a result of several later conferences between ourselves and the municipal power division, the city dropped that portion of its suit related to cable damage. The balance of the claims were also settled on a reasonable basis satisfactory to both parties.

The Case of the Unhelpful Consultant

A few years ago while my cellist wife was attending a string teachers' music workshop at a university in the south of England, I went along, hoping to play my flute a bit, but also planning to look into some of the English research work going on in electric power transmission. I had written to several of the large power equipment manufacturers who maintained notable research laboratories and to the Central Electricity Generating Board for information on visiting these facilities. I received very cordial responses.

When I returned to the university town where the music workshop was in progress, I found I could indeed get in a little playing myself. Although this was strictly a string workshop and my wife was enjoying the cello lectures and master classes, the workshop did put on several chamber orchestra concerts and found my flute playing and the playing of a couple of other nonparticipants, who played French horn and clarinet, very useful.

During the second week of our stay my pleasantly lazy music idyll was interrupted one morning by a telephone call from an engineer with the Central Electricity Board. The conversation went like this:

C.E.B. engineer: Sorry to bother you, but we heard where you are staying attending that music workshop. We do have a bit of a problem, a failure on a new distribution cable circuit in the town there, and as a matter of fact, it has boiled down to a bit of controversy between the cable supplier and ourselves. In view of this and since we are dealing with a foreign supplier, we would like to engage you to look into this matter for us.

I: Thank you for your consideration. However, my being here with my wife is sort of a vacation for me. I have nothing with me in the way of references, text books, or even a calculator or slide rule. Also, we are planning to leave here in three or four days and spend some time in London before flying out.

C.E.B. engineer: We appreciate the position, but we believe you can settle this matter for us within the next day or two at most. The supplier's cable engineers are here now in our branch office in town. If you will telephone our branch manager, he will pick you up as soon as you can be ready. The matter is fairly urgent, and we hope you can be with us as soon as possible.

Of course I was glad to be called and in a matter of an hour or so was in conference with the branch manager and his engineers, who laid out the situation and their contentions as to responsibility for the cable problem. I then met the two Italian supplier's engineers and we agreed to all meet in the morning of the following day to iron out differences and arrive at a settlement.

Briefly, the situation was as follows:

The town's major substation was located at one end of the High Street and this was at virtually the highest elevation of the town. From there the high street dropped down quite steeply at what looked like a 10 percent slope for about one and a half miles to a river that bordered the west end of the town. Under the street there were a number of cable ducts carrying various moderate voltage cable circuits down to another substation at the river's edge.

About two years ago, anticipating the need for greater power requirements in the industrial section beyond the river, the local utility had installed a new thirty-three-thousand-volt three-phase cable circuit under the High Street and then ran it under the river as submarine cable. This consisted of four lead-sheathed (one being a spare) solid type cables containing the current-carrying conductors, each insulated with oil-impregnated paper tapes. The term *solid type* indicates the cable is completely self-contained and for this moderate voltage does not require internal oil flow and an external oil pressurizing means.

Well, as it happened, after two years of use, two of these relatively new cables failed and the whole circuit had to be taken out of service just a couple of weeks before I arrived on the scene. When the first failure occurred, the utility had promptly connected the spare cable as replacement and resumed service—but not for long. Ten days later another of the cables failed.

In view of the short operational period and the absence of any external damaging influences, the utility board had taken the position that all of the cables as delivered were defective and that the usual one-year warranty was really not pertinent to an impregnated paper cable for which much experience has indicated satisfactory service lives of thirty, forty, or even more years. Therefore, they had called in the cable's manufacturer and demanded that he replace all the cable lengths and carry the costs of removal as well as installation of the replacements—plus lost revenues during the outage time.

The manufacturer's sales representative had examined the cables shortly after the first failure and completely disagreed with the utility. They claimed that their cables were in no way defective and that the utility's own operational practice was the cause of the failure. This then was the impasse that the utility's engineers, the newly arrived manufacturer's engineers, and I had to resolve during the next twenty-four hours.

For the rest of that day in borrowed overalls I climbed in and out of manholes on that steep High Street. I was able to examine the failure regions of the two cables, since the street had been dug up for some distance after they had been located. Both failures occurred only a few hundred feet below the upper substation and were about twelve feet apart.

This was very significant and its implications sent me scurrying down to the terminal vault in the river substation. Sure enough, I found what I suspected: All of the cable lead sheaths were greatly distended and in the case of the two cables that failed the lead had burst open along several crack lines and oil was seeping out very slowly but continuously. It was obvious that these burst sheaths were due to excessive internal mechanical pressure, not burnouts. It was also obvious that the cable oil, although a tacky variety, had during the two years of installations migrated downward and built up a high static pressure at the base of the run. This so depleted the oil impregnation of the insulation at the top of the run that the cables could no longer sustain the service voltage there and failures were inevitable.

After seeing the burst sheaths, I went back into the three manholes along the High Street and very carefully examined each of the joints there. I was looking for stop-joints, which are normally used to isolate the built-up static pressure head of the oil contained in the cable's insulation on long descents.

An unrestrained column of oil creates some four pounds per square inch of pressure for every ten feet of height. In this case the High Street length was 6,500 feet at a 10 percent slope, which produced a vertical oil column 650 feet in length and thus a static oil pressure of 260 pounds per square inch at the bottom. These cables had double lead sheaths and an overlaying wrap of Hessian tape, but this arrangement at most could withstand a working pressure of only eighty pounds per square inch.

Well, my finding in the joint manholes was exactly what I expected: all of the joints were "through" type. There wasn't a single stop-joint in the whole run. It was evident that this cable circuit had been designed to have through stop-joints, breaking up the run lengths into four 1,625-foot sections, each with a maximum static oil pressure of only sixty-five pounds per square inch. However, somebody goofed and put in through-joints, with the consequences as found. As they tried to explain it later, this was the first oil-impregnated paper cable installed in the town. Their other distribution primary lines were at only eleven hundred volts, and these cables were insulated with varnished cambric tapes containing no fluid oil.

I reported my sad news to the utility management late that afternoon. Next morning we all met, supplier engineers and ourselves, and the matter was settled in a few minutes. The Italians were all smiles; our side looked pretty glum as we made our admissions of fact.

And so, as you can see, I didn't help my English client very much.

The Case of the Hanging Hazard

It happened on a county road somewhere in Idaho. The east-west interstate had some bridge repair work in progress, and for about two weeks traffic was being detoured onto two almost parallel county roads. The northerly one was designated for westbound and the southerly one for eastbound vehicles. Temporary exits onto the detours and back to the interstate were abrupt but well marked. The southerly detour road had a wood-pole transmission line running along its length supporting a sixty-nine-thousand-volt, bare-wire, three-phase circuit and some lead-covered telephone cables lower down. The poles were spaced approximately four hundred feet apart and were placed about three feet off the narrow road's shoulder.

According to a trucker who witnessed the accident, shortly after midnight a racing type sports car with two occupants came barreling around the east indicated detour exit at a very high speed—so high, in fact, that the car didn't or couldn't get straightened out onto the county road and smashed into one of the power poles.

The force of the collision was so great that the pole was splintered and snapped about three feet from the ground. Both occupants were thrown out of the car and injured severely. The trucker using his C.B. transmitter was able to alert the state patrol, and in less than twenty minutes several patrol cars arrived and shortly after an ambulance came. One of the occupants, the passenger, I believe, was thrown to just off the road shoulder. He was picked up on to a stretcher and quickly placed in the ambulance. The driver of the car had been thrown well beyond the shoulder into the undergrowth that bordered the road. One of the officers using his flashlight spotted this injured man, and the officer and the two stretcher bearers hurried into the undergrowth and picked him up. They apparently took a diagonal route back to the road, the officer leading with his light. And so they walked right into the downed transmission line wires *still energized* and clearing the ground at that point by some eighteen inches. The patrol officer was instantly electrocuted. The lead stretcher bearer was burned very badly and later succumbed. The injured car driver on the stretcher and the rear bearer essentially escaped serious injury. Each wire of the downed line was at forty thousand volts to ground.

The power company, owners of the transmission pole line, was sued by the families of the men who died in the accident and also by the second stretcher bearer. Without getting into too much of the legal details of these suits, it appears that the essential complaint was that the power company failed to instantly deenergize the transmission wires when the pole was broken and the line came down. It was repeatedly stated that any modern electric power system, with its many safety devices for almost instantly interrupting power flow and the telecommunication system going over the same lines for voice, signals, and sensing functions, could be able to tell and act when one of its lines went down.

Power engineers in their depositions explained that had the line come down in actual contact with the ground or in such close proximity as to arc over to ground or to low moist undergrowth, the ground relays at the controlling substation would have responded and tripped out the line circuit breakers. Since the downed wires were well off the ground, no ground current could develop and the relays would not operate. However the plaintiffs' attorneys would not buy this explanation and proceeded to seek a trial date.

Idaho is a rather remote state with a relatively small population, mostly in one large city. Most of the power company's transmission involved long rural lines. While the company was certainly a capable utility with many years of highly successful experience in power supply and modern up-to-date equipment, they raised their own question: "Do we have or even know of the latest state-of-the-art instrumentation available that can sense overhead line mishaps such as occurred in this case?"

Well, they organized a quick inquiry amongst neighboring power utilities and several of the national electric power associations, but nothing definitive resulted. At that point I was called in by the company and asked to make a comprehensive survey, worldwide, of state-of-art line-sensing instrumentation.

I was pretty well assured, in my own mind, that there was no way to tell if a small part of a transmission line was at installation height or at some other height clear of ground. With a span or two of the line down in the manner of the accident, the only resulting electrical change would be a slight increase of capacitance to ground only for that portion downed. In several thousands of feet of line this slight capacitance change (extremely difficult to measure alone) would be washed out completely. However, I undertook to make the survey.

It took me more than a month of inquiry in libraries and by correspondence and telephone to obtain 100 percent agreement that no sensing equipment was currently available anywhere in the world that could have deenergized that downed line.

The power company accepted my report, and with a clear conscience that it could not have avoided those fatalities, they nevertheless decided to negotiate settlements with the several plaintiffs.

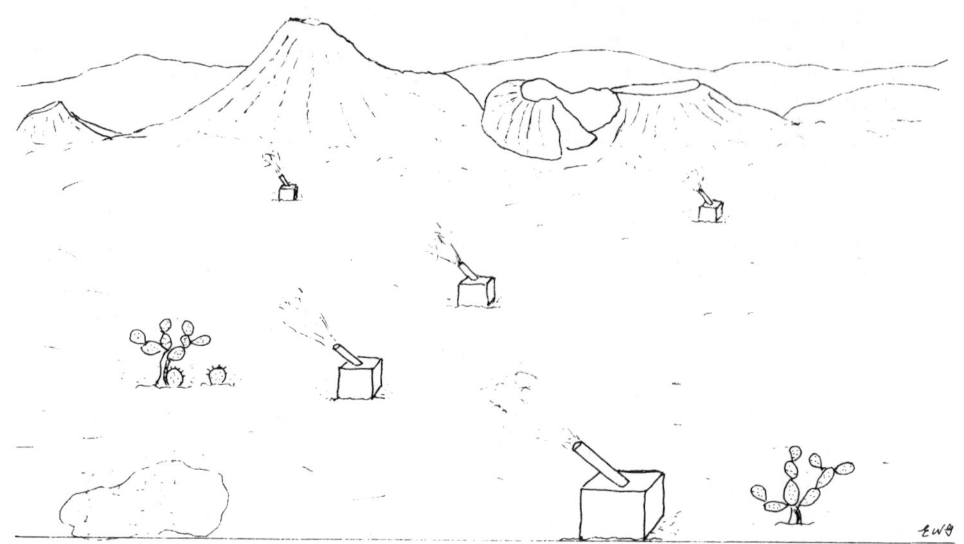

The Case of the Steaming Wells

Not more than fifteen miles south of the city of San Diego is the Mexican border and the entrance of Baja, California. This exotic piece of land is almost completely separated from mainland Mexico by the Gulf of California. It stretches south for some eight hundred miles. At the border, the Baja is almost 160 miles in width, but in most of its extent to the south it narrows to less than 25 miles. In its fifty-five thousand square miles of area there are deserts, ten-thousand-foot-high mountains, lush ocean beaches, and unsurpassed game fishing in its long gulf—into which our Colorado River finally empties.

The mountain ranges in Baja are evidently extensions of our Pacific Coast ranges, and like them they are of volcanic origin. In particular, within fifty miles of our border there was once an enormously active volcano chain known as Cerro Prieta. It is now virtually extinct. Except for some smoke seeping between basalt crevices here and there and the heavy smell of sulphurous chemicals, no action of gas or lava has taken place for several centuries.

However, lying deep beneath the surface in this area is evidence of a great pool of magma, molten rock, left over from the past volcanic actions. It gives ample evidence of itself in the many steam wells and hot pools that abound there. Some of these steam wells are amongst the largest in the world and have the highest temperatures and pressures.

There was, however, very little attention paid to these wells by the Mexican government other than to rope them off for safety and use them as fascinating exhibits for tourists. However, early in the 1960s these steam wells assumed a new importance. In early years Baja, California, was very lightly populated, and even as late as 1940 there were fewer than one hundred thousand inhabitants. After World War II, the Baja population began to build up rapidly in the cities, towns, and villages where some industries were started, but more particularly due to heavy funding for tourist attractions and facilities. Formerly sleepy towns like Tijuana, Ensenada, Santo Tomás, and San Filipe became awake with big hotels, bulging shops, factories, and warehouses. And by the late 1950s it was very evident that the old electric-generating facilities could not be pushed farther and some extensive new electric energy sources were urgently needed. By that time, the Baja population exceeded 1 million and it was still growing.

The energy planners of Mexico were well aware of a successful American steam well electric generating plant at Thermal, California, which had been operating for some time. With the much larger, hotter wells in Baja, it appeared to the planners that this might be a heavensent opportunity to provide the needed bulk electric power while at the same time doing this on a no-cost-for-fuel basis.

The government moved rapidly to implement the plan. The first task was to cap a number of the largest wells and record pressure and temperature build-up as basic data for design of a steam/electric generating plant. But capping a mighty, roaring steam well is no task for amateurs. At one well I saw later, a four-inch-diameter pipe had been driven into the well and a solid mass of steam was roaring out of it. At the suggestion of one of the Mexican geologists with our party, I picked up a rock weighing at least five pounds and threw it into the venting steam. Instantly the rock was shot out and high like a cannonball for more than one hundred yards. On another occasion during a later construction phase, a boom truck operator inadvertently allowed a three-quarter-inch steel cable hanging from the boom to swing into the venting steam opening pipe of a large well. The cable went flying in two pieces.

The Mexicans could not cap their well successfully. Finally they called an American company whose sole specialty was capping gusher oil and gas wells. That did it, and four or five steam wells were capped and under control—but only after several failures at each site. Temperatures and pressures at the capped wells generally showed superheated steam at approximately four hundred degrees Fahrenheit and pressures between 325 and 375 pounds per square inch.

Of course, the well temperatures and pressure are decidedly lower than required for a large, high fuel-efficient steam-driven plant using gas or oil. However, excellent generating plants can be designed to operate at these lower steam characteristics and fuel efficiency is no problem.

The Mexican government decided to go ahead and sent out bids for a one-thousand-megawatt generating plant utilizing the steam wells in number as required. In due time tenders were received from American, British, Italian, and Japanese companies, the last being the lowest cost bidder by a sizeable margin.

Facing up to the already heavy well capping cost and the projected generating plant costs, the government took careful stock of exactly what they would be getting for their money. And that brought up the question of water—water for the steaming wells. What was the source of the water coming up as steam? Was there a sufficient body of ground water to maintain the wells for the foreseeable future? They had no data on these steam wells going dry, but suppose the source of water was a nonreplenished subterranean pool or reservoir of limited extent whose runoff was reaching the hot magma. Such a reservoir might start to trickle out in ten years or so—and the generating plant would become useless. Their geologist warned that this could be a possibility and strongly recommended that a survey of the water source be made before going ahead.

At that time I was directing an engineering research division at a large northwestern university. Two of our research sections, Water Resources and Geology, had been teaming up in some very successful work on subterannean water identification and analyzing the extent of aquifers for city water supplies. Some of this work had been presented at various scientific and municipal engineering meetings, and several papers had been published in water resource journals.

It was not surprising, therefore, when the Mexican government contacted the division and asked me if we could undertake to identify the source and extent of the Baja steam well water. It was an interesting, important project, and we agreed to undertake the study. A contract was signed detailing sponsorship by the Mexican Geologic Survey and providing headquarters for our work at the city of Mexicali, just a few miles below the border.

In our studies on water resources up to this time we had had good success with a two-prong approach. First a thorough geologic survey is made of the area under study, cataloging all surface waters as to source, direction, rate of flow, and chemical composition. In addition, land features, old wells, rainfall, soil percolation, and ground water levels are checked and recorded.

Based on these data, several of the most likely surface waters are selected for carrying a tracer. The tracer is a short-lived radioactive isotope element, a very small quantity of which is dumped into the selected stream. Being radioactive, the tracer can be followed readily by a radiation-sensitive counter. Samples of water from the city's old or projected deep well drill holes are brought up and checked for the tracer. Generally, one or more of the surface waters inoculated will be identified and found as replenishing the deep well. If none of the surface or rain waters can be traced to the projected well, we can conclude that its subterannean reservoir is fossil water and has no replenishment.

Although our field truck was licensed, approved, and marked to carry radioactive isotope, it turned out that the Atomic Energy Commission would not approve transport of the truck through the several southern states and across the border. With this restriction, never previously imposed on our work, we feverishly explored another tracer technique known as Neutron Activation Analysis, or NAA. By this method a nonradioactive element that later can readily be converted to its radioactive isotope by suitable neutron bombardment is used as the tracer. Another requirement is that the tracer element must not normally be found in the waters to be traced.

We had been given a chemical analysis of the steam well effluent. We were then able to select several suitable tracer elements and made these up into the proper dosages to be dumped into the various Baja surface waters. Once this was done, we would take samples of the steam well vapor, condense the vapor to water, bombard the sample with our portable neutron source, and then check for tracer presence with a high sensitivity radiation detector. Mocked up experiments in the laboratory showed the method to be very effective. And this is what we took to Mexico.

Our field team consisted of the section heads for Water Resources and Geology and three technical assistants. In addition, the Mexican Geologic Survey had several of its staff there to assist the study. Fortunately, these latter were bilingual and were of great help. They met us when our two equipment-filled trucks arrived in Mexicali and helped in getting our personnel settled in suitable living quarters and finding working office and laboratory space. The setup thus appeared to be excellent, and we were ready to go to work and anxious to get started. But fate in the form of the Mexican Department of Labor decreed otherwise.

On the day following their on-site arrival, I had a telephone call from Jim, our head geologist, speaking from Mexicali. In an unusually high-pitched, frantic voice he was saying what I couldn't understand at first.

"Calm down, Jim," I said. "I can't understand a word."

He did, and in more measured tones but still filled with emotion he said, "Gene, we are all under arrest here. You have to do something right away. We are not allowed to leave the hotel, and I don't know what has happened to our trucks."

It was an incredible situation as Jim unfolded the story. Just as our party, including the Mexican collaborators, was getting the truck equipment unloaded, two uniformed federal policemen arrived. They immediately singled out our section heads and asked who owned the trucks and for what purpose they were here in Mexico. Our Mexican friends were as surprised as we were but answered politely that an American university had been contracted by the Mexican government to carry out a geologic survey in the thermal area of northern Baja and that work was now ready to begin. With that the policemen were all smiles and said all they needed to do was to stamp our party's work permits and they would be off. But we had no work permits; nowhere in our conferences and contract with the Mexican officials had work permits come up or even been mentioned.

"But you must have work permits for each of your party," the older of the policemen said. "No foreigner may work for gain in any manner in Mexico without a work permit. We cannot permit you to engage in whatever you planned to do without these work permits."

The upshot was that our party was placed under house arrest and the trucks impounded until work permits could be produced. Our Mexican collaborators assured our party that as soon as they got in touch with their geologic chief in Mexico City the problem would be resolved speedily. They did early next morning—but the problem was not resolved. It only got more complex. Each department accused the other of failing to follow proper procedures, and in particular, the Labor Department refused to give an inch. They insisted that work permits, properly applied for and approved, had to be obtained and produced. And they had the authority apparently to back up their demand.

This was bad news, but worse was yet to come.

It appeared that the proper procedure for a foreigner to obtain a work permit was to go to the closest Mexican consulate in his country and make application there. If a proper need to work in Mexico could be shown and if this work was for the benefit of Mexico, et cetera, a work permit would be issued by the consulate. So when Jim called me about their house arrest and the further complications, he said that I should immediately contact the Mexican consulate in Seattle, file applications for the party, and send the work permits to them in Mexicali as quickly as possible.

Armed with a copy of the water survey contract and personnel data on each of our party, I drove to Seattle, called the Mexican consulate, and got an early appointment to see the consul. She was very cordial and clucked several times when I told her our predicament.

"Well," she said, "we can settle that matter easily. I will approve and issue your work permits for the survey. Just have each of your party come to the consulate here for his permit, photo, and signature. It won't take more than ten minutes for each one."

"But, Madam," I protested "these men are right now actually under house arrest in Mexicali and under close surveillance by Labor Department marshals. They cannot come here."

She looked blank for a moment and then with a sad smile said, "In that case, I can do nothing for you. Work permits must be issued in person to the person." And that was that.

I called Jim and told him that the consul route wouldn't work and that I was going to try to make contact with the U.S. ambassador in Mexico for help. After much inquiry at our State Department and considerable telephone delay, I did reach our ambassador finally. He listened quietly to my concerned voice, and I outlined what had happened.

"This is not unusual down here," he said. "It can happen anytime. However, I think I can clear up the matter, but it may take a day or two."

And he did. On the fourth day of the confinement, the same two federal policemen arrived in Mexicali and with wide grins on their faces handed our work permits to each of our party and then stamped them after the signatures were affixed. The house arrest was over, and we were free to go to work at last.

After that traumatic experience, the survey got under way and continued more or less smoothly for the next several months. We held many telephone conferences between my office at the university and the field party to iron out minor difficulties and establish priorities. At one point it seemed best for me to visit the site for a few days. (I wanted an excuse to do this anyway.) So, taking my wife along, I traveled down to Mexicali and spent some time looking over the survey results but, more pleasurably, touring the entire Baja thermal area and inspecting the capped and uncapped steam wells. I was particularly impressed by the great quantities of caked salts of many shapes and colors that surrounded the wells and even the hot pools. I suggested that the salt deposits be analyzed for composition, since they might be quite valuable for agricultural soil enrichment.

We completed the survey and had a report in the hands of the Mexican government in just two and one half months from day one. Our report showed that the water for all of the steam wells tested came from ground water pools reaching the hot magma and that these pools were constantly being replenished by

Colorado River water. Our results were so consistent in every respect that there could be no doubt as the accuracy of the conclusion.

This story has a sequel, the details of which I learned later from my Mexican engineering friends. A generating plant was built by a Japanese company utilizing the output of a number of capped steam wells. It was placed into successful operation and started supplying the long needed power over some new transmission and distribution systems. After a few months of operation, however, the plant had to shut down due to excess salt formation in the steam separator that caused continuous clogging of the inlet steam feeders. The function of the steam separator, as its name implies, is to keep any salts from entering the inlet steam lines. The Japanese-designed modifications were made by them, but none was effective. Finally an American company was called whose experience in building successful steam separators for the world market was well known. They solved the problem and the plant went back into production. As far as I know, this no. 1 steam well generating plant is still going strong.

An important side benefit to the Baja agricultural economy has been the copious separated salts containing phosphorus, potassium, magnesium, sulphur, and other valuable soil rebuilders.

The Case of the Deadly Bight

A seaman's word from the day of sails, *bight* still means that portion of a slack rope or line between its ends. And one of the important rules a sailor learned quickly when he went to sea was "never stand in the bight of a rope." If someone is unfortunate enough to be within the bight or inner curvature of a slack anchor chain, for example, and the anchor is suddenly let go the slack straightens out almost instantly and with such great force that can seriously injure or even kill. That is what this story is all about, and though it deals with power transmission lines and linemen rather than ropes and chains and seamen, the old rule still holds: Never stand in the bight of a line.

A municipal power company was installing a long needed high voltage overhead transmission line spanning a multifreeway complex. The steel towers, set between the roadways, had to be abnormally high so that no part of the line could possibly interfere with or endanger the heavy traffic below. The power line to be strung on the towers was a thick, bare cable composed of a number of layers of aluminum strands surrounding a core of steel wires. When completed, six of these cables, representing two three-phase circuits, would be held by suitable insulator strings on the tower crossarms. Each cable would be under a predetermined, rather high tension so as to yield a required maximum sag under all operating conditions and weather.

Let me sketch very briefly the essential procedures for putting such a line up, since it is pertinent to this story. The cable is supplied wound on reels that are set up on reel mounts permitting each to turn freely. The mounted reels are placed in a special reel trailer, which incorporates braking mechanisms for controlling the rate of payout of cable from the reels. A payout lineman constantly attends to this control. The cable coming from the trailer passes through a line-tensioning machine also under constant operation control and then up to the first tower. At this stage, the cable is passed through a temporary sheave (a grooved wheel) and then attached to a steel pulling line, which pulls it through sheaves at each tower until finally, after the last tower, it is passed down again through another tensioning machine at ground level. During all of this threading of the cable through the towers, the braking action at the payout reels plus the pulling action of tensioning machines at either end of the line must act in perfect coordination to provide the cable with the required stringing tension. When this operation is completed satisfactorily for all six cables, they are taken out of the sheaves and clamped to the bottom of the already hanging suspension insulators and the far ends are permanently fastened to dead-end insulators.

In the above I have not described how a line-tensioning machine functions. It is a massive piece of machinery with several heavy duty motors operating several capstans about which there are one or more cable turns. Wherever the cable moves in, around, or out of the tensioner there are large coated rollers to prevent scoring or squeezing. At the point where the cable first enters the machine, it passes through a split hardwood die, mounted on a free swinging arm, whose opening is set at a fraction larger than the cable diameter. Purpose of the die is to "iron down" any of the outer strands of the cable that may be loose and tending to "bird cage" outward. The cable then takes a single turn over the first motor-operated grooved capstan and passes over rollers then up to the second capstan and from there goes through an adjustable hardwood throat aimed at the first tower crossarm. The second tensioning machine, located at the far end of the tower line, does the pulling at the selected tension by means of its motor-operated capstan, and its steel pulling line is wound on a storage drum as it comes in. A continuously reading tension meter is located in this machine.

As indicated previously, the stringing tension is set at such a value that the midpoint sag of the cable between tower supports will not exceed a specified value. The tension to accomplish this depends on many factors: the size, weight, and type of cable, span length between successive supports, ambient temperature range, and expected weather conditions. Should these factors, when considered together, require the worst condition tension to rise to more than one-third of the cable's breaking strength, either a stronger cable must be used or the projected tower or pole spacing will have to be reduced.

Now back to that new transmission line being installed over the freeways in a busy west coast city. The reel trailer with three full cable reels in place was positioned about eighty feet from the first tensioning machine. The cable from one of the reels was carried forward to the first tensioner, threaded through the die, taken around the capstans, and pulled a short distance out of the aiming throat. The temporary sheaves on the towers were all in place, and the steel pulling line from the second tensioner had been threaded through them and then down to the cable waiting at the first tensioner. Here the pulling line was clamped to the cable using a suitable compression sleeve. The breaking strength of the cable was thirty-two thousand pounds, and the calculated worst condition that could produce the maximum permitted sag required a tension of nine thousand pounds. Since this was well under one-third of the cable breaking strength, that was the basis for setting the cable tension during each day's stringing operations. For conditions on the day of the accident this came out to seventy-five hundred pounds and therefore the coordinated controls in the cable payout reel and in the two tensioning machines were set to yield this amount of tension in the cable.

Before starting the pulling, two linemen were stationed along the length between the payout reel and the first tensioner. They wore heavy gloves and their job was to keep hands on the moving cable to detect any abnormality such as a popped-up strand or loose outer strands or any significant surface damage. For no apparent reason both men choose to stand on the side of the cable toward the takeoff to the transmission towers. No one present, including the crew foreman, appeared concerned at the position these two linemen had taken.

By midafternoon all was in readiness and the cable pulling began. They brought the tension up to the seventy-five-thousand-pound setting quite easily, and this was being held satisfactorily through the sheaves on the first three towers. And then it happened! Suddenly the taut cable over the lead-in capstan in the first tensioner jumped out of its groove and at the same time a powerful whiplash wave ran back along the cable stretching between the payout reel and tensioner. This caught both linemen standing on that side of the whipping. It struck them with such force that one was thrown almost thirty feet and the other ten or twelve feet. The man receiving the hardest blow landed head first against a stack of steel tower angles piled on the ground. He never regained consciousness and died on the way to the hospital. The other lineman was badly injured and was hospitalized more than a year before being released, only partially recovered.

Reconstructing the tragedy was not difficult. When the cable ran off the three-foot-diameter capstan, it instantly put that much slack in the cable—which was immediately snapped out by the pulling tensioner. That caused the whiplash along the cable where the linemen were standing. And where they were standing was on that side toward which the cable was being snapped straight. In other words, they were in the *bight* of the cable.

The families of these men entered suit against the manufacturers of the tensioning machine, the payout reel trailer, and particularly the cable manufacturer. They could not sue the power company because they were employees and state workman compensation laws, to which the company subscribed, prohibited that. The complaint against the cable manufacturer was more serious. It contended that the outer strand layer of the cable was wound very loosely and that there was some kind of obstruction on the inside of one reel flange that gouged the cable as an end turn was paying out. This, the complaint stated, caused the outer strands to bird cage outward several inches so that when that portion of the cable entered the tensioner's capstan, the bird cage lifted the cable out of the groove and caused the accident.

The attorneys for the cable manufacturer called on me to investigate what had actually happened during the line stringing operation and in particular what part, if any, their reel and cable may have played in precipitating the tragedy. During my investigation I was able to inspect the tensioning machines and the reel trailer as well as the cable and its reel. Every foot of the cable that had paid out of the reel up to the time of the accident had been cut off and stretched out on the ground in a large field. Therefore, I was able to examine its condition minutely. With regards to the reel, I did find a small eye-bolt bolted to the inside of one of its flanges. It was located about six inches below the top of the flange at approximately where the fourth layer of cable on a full reel would come. The reel in question had been full.

Knowing the exact distance between the payout reel and the tensioner capstan and having all of the cable that had been paid out now stretched out, I was able to examine where the bird caging was supposed to have been started by the eye-bolt and where the supposed bird caged cable had run off the capstan. I found some superficial gouges here and there on the strands but no signs of any bird caging. Another item of interest developed when I calculated exactly where the eye-bolt was located with reference to the amount of cable that had paid out. This rather simple calculation, since I knew the full length of the cable and its diameter and all of the reel dimensions, definitely showed that the eye-bolt could not have come in contact with any part of the cable that was paid out. In fact, the eye-bolt was located two layers below the last layer that had been paid out up to the time of the accident. In other words, the eye-bolt was adjacent to cable that had never left the reel.

However, the fact that for some unknown reason the cable company had left an eye-bolt mounted on the inside of the flange where it could gouge a cable was made much of. There was another eye-bolt mounted on the outside of the reel flange near its top, to which the end of the cable had been fastened properly. Although the cable company as a result of my investigation was able to disclaim any responsibility in the accident, its attorneys talked them into negotiating a settlement with the plaintiffs.

My own feeling was that the cable company, as defendant, should have permitted its case to come to trial. There was a most important lesson to be learned on how this accident occurred and how it could have been prevented. By negotiating a settlement between the attorneys of the parties in a closed room these lessons would not be learned.

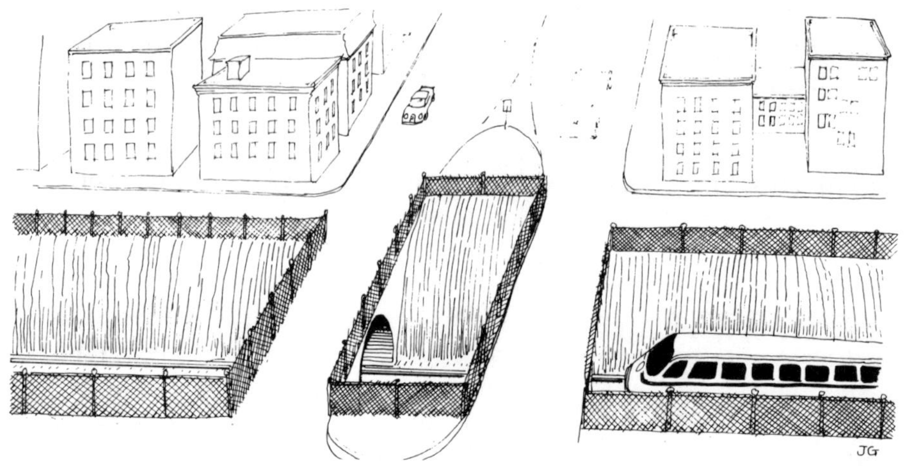

The Case of the Wooden Lifesaver

One of the most imaginative and successful mass transit enterprises has been the Bay Area Rapid Transit system, more affectionately termed BART. There is no doubt that it has proven itself of enormous benefit to Bay residents and commuters between San Francisco and surrounding cities and towns. The trains are usually made up of two articulated cars and are operated either fully automatic or, if present, a motorman can override the automatics and take control. Everywhere in the automated system are built-in safety provisions, which can sense a right-of-way problem or a rail or car damage and then set up the necessary protective action to be taken.

The propulsion motors on the cars are designed to operate on direct current (dc) at one thousand volts. Power is picked up by a special contacting shoe, which extends out at near track level and rides on a third rail energized at one thousand volts. Since the third rail must carry the current required by all the operating trains in a given section, it must be made of fairly low resistance material, yet be strong enough to withstand installation on the ties and pickup shoe pressures. As finalized, the third rail is shaped and sized like an ordinary train rail, but it consists of two metals. The mounting plate and core of the rail is steel. Surrounding the steel core is a very thick layer of electrical conducting aluminum, which is bonded to the steel. It is this aluminum layer that is contacted by the car's pickup shoe and carries 90 percent of the car currents. Since the third rail is at one thousand volts to ground, it is not mounted directly on the wooden crossties but is supported on strong ceramic insulators having adequate flashover strength under normal weather conditions.

When BART enters a town or city, it usually proceeds to a loading station by dropping into a tunnel under the streets. The tunnel usually has both open and completely closed-up sections. Where the tunnel top is open, the third or "hot" rail is exposed. And as such, it is subjected to atmospheric contamination and to trash or garbage thrown from the street level. Also, it constitutes a great danger to children who may have climbed down into the tunnel for a lark or to anyone walking the tunnels for any reason. During the first year or so of BART's operation there were several suicides and near suicides involving the energized third rail.

Therefore, it became obvious that BART needed a cover over this "hot" rail, certainly in the open tunnel section and perhaps in other sections of the line. A satisfactory cover had to provide continuous, good electrical isolation from the energized rail, be strong enough to support several hundred pounds dropped upon it, and in no way interfere with operation of the trains' power pickup shoe. Also, due to the very large footage required, the cover had to be economical both as to material, installation, and maintenance.

One of the electrical manufacturing companies located in the Bay Area became interested in bidding on BART's third rail cover. I had some previous consulting work with this company on a motor insulation problem that turned out successfully, so they called me to have a look at BART's requirements and to assist them in a preliminary cover design. Our solution, the one that won the bid, turned out to be very simple: a properly located hardwood cover, mounted on reinforced plastic brackets, both easily replaceable, with the whole shaped to shed water, trash, and bodies. These cover assemblies, as far as I know, are still in service protecting the third rail; the wood covers have been replaced from time to time.

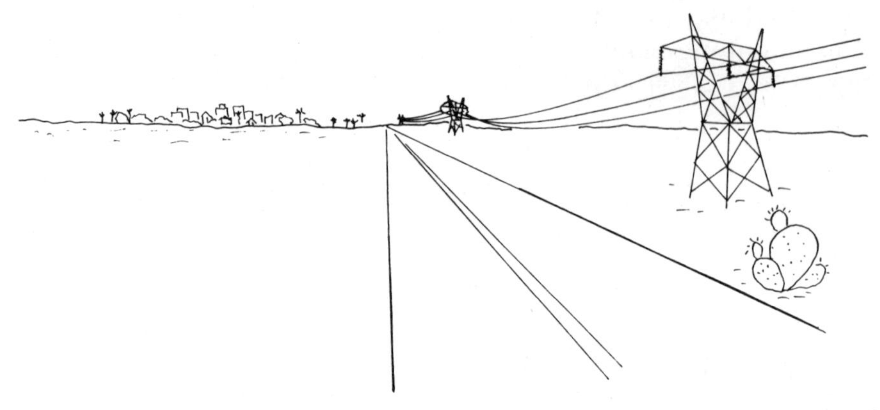

The Case of the Worried Mayor

One late spring a few years ago when my wife and I had just returned from a three-week stay in Great Britain, I found a number of calls logged on our new telephone-answering gadget. Most of these were local, with some of the callers having hung up without leaving a message. However, there were three calls from the same person spaced, as near as could be judged, about a week apart. In each of these calls the caller simply identified himself and asked me to return the call as soon as possible.

I did return the call that same afternoon and then listened to an interesting story and a subsequent request for assistance. The person I was talking to was an aide to the mayor of a large city in Southern California. It was the mayor himself who had asked his aide to call me about a problem concerning an electric power transmission project.

It seems that the city's municipal power branch was planning to bring a large block of badly needed power into the city from a major generating plant in Arizona, some 260 miles away. The selected route for the transmission line was to zigzag through a dozen California towns and community projects, each of which would tap off a share of the energy for its own use. As planned this major power link was to carry nearly 2 million KVA at a very high five-hundred-thousand-volt line potential. Also, as planned this was to be a double-circuit three-phase, alternating current overhead transmission line on high steel towers.

When the city's plan to provide this additional power became known generally, there arose a storm of protests from the en route towns to be served, from farmer groups, from individuals living along the proposed right-of-way, from environmental groups, and from county political agencies and clubs. None of these protests concerned design or technical aspects of the proposed line. All of the complaints were largely based on aesthetics. The beautiful California countryside, they said, would be defaced if hundreds of these tall, ugly steel towers with hanging conductors marched across the land. While need for the additional power was conceded, the protesters insisted that an overhead line was not the only solution. "Why not put the transmission underground?" they said, and that was the proper solution.

Conferences between the mayor's office and his electric power department engineers did not get very far in resolving the matter. These engineers, working for one of the largest municipal power systems in the country, flatly stated that an underground system of the magnitude required would be impossibly expensive, to say the least.

Here then was the mayor, with his political eye on the governor's office someday, caught between the "devil and the deep blue sea"—a very uncomfortable position. As a result, he decided to get some unbiased outside expert opinions before making any judgments or commitments. But that turned out to be troublesome, too, for when he announced his decision to seek outside consultation, his own engineer group as well as the dissenters' acting group each wanted to select the consultant. So it became a merry-go-round to find a suitable expert acceptable to both factions.

I understand this stalemate continued for a few weeks until my name came up for consideration. Perhaps by this time the guessing game had become tiresome, so when the city engineers proposed my name as a long experienced, bona fida transmission wire and cable engineer from out of the state, the dissenting group agreed.

Once I learned the elements of the proposed transmission, I knew that the city engineers were right in their contention before I started to review the voluminous background material. But the case for undergrounding was even worse than they had anticipated. For one thing, underground transmission of that circuit length was not feasible using alternating current. Also five-hundred-thousand-volt underground cable technology had, at that time, not yet emerged from the laboratory. Various types of high fluid pressure-impregnated paper-insulated cable were under experiment using newly developed insulating fluids and plastic-coated paper tapes. But no commercially warranted cable was yet available nor could it be foreseen "just around the corner."

To transmit the required power at the then highest potential level for which direct current cable was obtainable namely 350,000 volts, would require twelve parallel cables, each with a copper conductor approximately one and one-half inches in diameter and an overall diameter of six or seven inches. Costs for 260 miles of such an underground cable system including joints and terminations, rectifying and inverting equipment, pressurizing and monitoring apparatus, et cetera, would be fantastic. Also, a direct current line does not lend itself to tapoffs, such as envisioned for this transmission. Comparing the original overhead five-hundred-thousand volt alternating current system as proposed by the city engineers with a possible underground direct-current cable line, the latter could conceivably cost several hundred times more and be years in construction.

My report, containing detailed computations but with conclusions essentially as above, was in the mayor's hands within a week of the time I was engaged in the controversy. The report was accepted and acted upon. The new Arizona/California transmission line was completed early in 1981. It is an overhead line.

The Case of the Sleepy Physician

Houston was and is a booming Texas town, so much so that its layout to a stranger from more conventional parts is almost incomprehensible. What was once called a downtown is lost amongst great cement- and glass-sided office buildings and huge convention centers. All of the shopping is done in so-called malls located miles out on roads radiating in every direction of the compass. Along these radial highways large residential developments are scattered, mostly apartment and condominium complexes. No place can be reached by walking; indeed, there is in fact no place to walk—the auto rules supreme.

I was called by a well-known insurance company to visit there in connection with a fire that occurred in one of these large apartment complexes. Although the fire gutted only one of the detached apartment units, the occupant lost his life and the state was investigating possible negligence or even criminal charges. The insurance company happened to be a major owner of this apartment group, and the particular reason they called on me, with my aluminum technology background, will soon be apparent.

From the voluminous information and depositions made available by my client I was able to fairly well piece together the elements and sequence of what happened. The resident of the apartment was a young, unmarried physician, Dr. Dee, who had recently become associated with a well-known Houston medical group. An older physician, also associated with that medical group and living in the same apartment complex, shared a ride pool with Dr. Dee.

On a weekday morning when the older physician, Dr. Jay, had the driving chore, Dr. Dee failed to show up. Although it was the first time that an unexplained absence had happened in more than a year of ride sharing, Dr. Jay thought nothing of it and went on to this office. But again the next morning Dr. Dee was not at their usual meeting place. Dr. Jay became uneasy and telephoned their offices at the medical group to ask if any messages by or for Dr. Dee had been received yesterday or that day. There was none. Thoroughly alarmed now, Dr. Jay ran over to Dr. Dee's apartment, rang and banged on the locked door, and then went to get the building custodian with his passkey.

Well, they got the door open and had just managed to step in when with a tremendous roar a great rush of flame erupted from an inner room. They were literally blown out of the door by the explosion, and some of their clothing caught fire. However, they were unharmed and could quickly beat out their burning clothes. By this time the apartment unit was well ablaze and a fire alarm had been turned in. The fire station responded quickly, but the interior of the building was well gutted before the fire was finally extinguished. And then they found Dr. Dee in the bedroom, lying on the charred bed dead.

The inquest into the fire and Dr. Dee's death included the fire marshal's investigation and a medical post-mortem and autopsy. The fire marshal issued a report stating that the fire had started in certain smoldering electrical receptacles due to the fact they were wired with aluminum conductors. The medical report stated that death was a result of the inhalation of massive amounts of carbon monoxide, not burns from the fire. It also gave as an opinion that death had occurred at least twelve hours before the burning of the building.

That was about all I had for background when I arrived in Houston and started looking into things. I was particularly interested in the fire marshal's report and was able to discuss it with him. He told me it was well known that wiring a house with aluminum was dangerous because it was not as good a conductor as copper and loosened at connections. He said that he had found some loose connections in the wall electrical receptacles of the burned apartment and once he realized they were wired with aluminum conductors he knew the cause of the fire. He played down any explanation of the fire blast when the door had been opened and also hinted that he might not agree with the medical report findings. I came away feeling that the fire marshal was a very opinionated man and it might be difficult for him to consider new facts of which he was not aware.

I spent half a day in the apartment studying the receptacles the fire marshal had pulled out, and I also pulled a few out myself. Of course the plastic parts were either gone or charred badly. Those receptacles that were badly burned showed loose wire connections, but in the few that were relatively unharmed the aluminum wire connections were still serviceable. None of the wire connections had serrated washers. None of the receptacles was of the type recommended for aluminum wiring.

This apartment complex apparently had been wired with aluminum conductors throughout about two years before the fire. I learned that a local electrical shop had installed all of the electrical service, so I called the shop and went out to talk with the owner. The conversation may be summarized as follows:

I: Why did you use aluminum wiring in the buildings?

He: Aluminum is much cheaper, lighter, and easier to install. It's approved by the National Electric Code, same as copper wiring.

I: Did you use any of the special connection devices for aluminum conductors such as receptacles with aluminum studs?

He: No. You don't have to use any of those things. We connect aluminum same as we do with copper—have been for years with no trouble at all.

I: What about using the serrated washers under the aluminum when connecting it to a copper stud?

He: No, we don't fool with that either. It's an extra operation that's unnecessary.

I: What about scraping off the oxide film on the aluminum wire before making the connection and putting on some antioxident grease?

He: If the aluminum surface looks whitish we'll scrape it; if not we don't. Grease? Wouldn't think of it.

I: Are you aware of the instruction tag for connecting aluminum wiring that accompanies every coil of aluminum building wire?

He: Sure. Those instructions might be okay, but they're not practical for us installers to follow and, as I said before, not necessary.

I came away from this interview feeling that the shop owner was quite disturbed by my questions, despite his hearty assurances to the contrary. That this guess was accurate was substantiated later. A few days after our talk, the electric shop sent several electricians out to the apartment complex to tighten all receptacle connections.

Before I go ahead with my story, let us pause here to set down a few facts about aluminum building wire. Aluminum has only a little more than 60 percent the conductance of copper. But that is more than compensated by standardizing on aluminum wire being two sizes larger than the correct copper size. So as far as current-carrying capability, aluminum wiring is equivalent of copper wiring. However, there is a problem at connections. Aluminum has a naturally formed, very thin coating of aluminum oxide. This is actually a ceramic material and is insulating. It is so thin that it will crack off under pressure and can be scraped off easily. Aluminum has a greater temperature expansion than copper and considerably lower tensile and compressive strengths. Also, aluminum wire is more subject to fatigue cracking at nicks. All of these factors should be taken into account in order to make good, permanent electrical connections with aluminum house wiring. Instructions to make good permanent connections have been publicized repeatedly in the technical press, in hand-out pamphlets, and on tags attached to the aluminum wire coils as sold. Many high grade electric shops follow these instructions; many do not. There have been fires resulting from incorrectly connected aluminum building wire, but, contrarywise, a great number of such incorrectly connected wiring has lasted for years without trouble of any kind.

Was a loosened aluminum wire connection in Dr. Dee's apartment responsible for what happened? Dr. Dee's sister believed it was and entered suit against the apartment's owners, the manufacturer of the aluminum building wire, and the local electric shop. Damages asked were well into seven figures.

I went back to the burned-out apartment a second time with one of the insurance company's young attorneys. I was looking for the loose connection receptacle that may have initiated the fire—assuming the fire marshal was right. After closely examining each possible receptacle box, I concluded either the aluminum wiring was not the fire origin or their charred condition was too great to be able to tell. It must be remembered that in order for a loose connection at a receptacle or other connection point to overheat, that receptacle has to have an electrical appliance or lamp, et cetera, plugged in and *turned on*. This limited the possibilities to those receptacles into which something is plugged and, in addition, that something is likely to be turned on for a period of time unattended—for example, lamps, (there were no ceiling fixtures), the refrigerator, and the bathroom wall lights. Since it was the middle of August, I eliminated the electric wall heaters.

I was about to take another look in the kitchen area when the young attorney called to me excitedly from the bathroom. There amidst the charred remains of a plastic-topped cabinet was a melted-down piece of metal that obviously was once a hot plate. Rooting around nearby, we found shards of Pyrex glass with melt-rounded edges that appeared to have been part of a teapot. With a little more effort we were able to identify a round, more deeply charred area in the plastic top of the once cabinet that approximately fitted the bottom section of the once hot plate. With that finding the probable sequence of events that fateful night started to unfold.

Two days before the fire in the evening, we know Dr. Dee had gone back to his office at the medical group and worked in the laboratory. He must have returned to his apartment late that evening, plugged in the hot plate in the bathroom, and put on a small glass teapot of water—probably for a nightcap of tea or instant coffee. It appears that he then threw himself on the bed for a stretch-out while waiting for the water to boil. Evidence of this was the fire marshal's report that when found Dr. Dee still had on some remnants of his street clothing. It was late, he had been working all day and evening, so it is not surprising that once down on his bed he fell into a deep slumber. He never awoke.

As he slept, the water boiled out of the teapot, the red-hot glass finally burst apart, and the pieces together with the red-hot hot plate started a smoldering fire on the plastic top of the cabinet. Eventually the electric cord of the hot plate must have burned apart, which cut off the hot plate's continuing heating. However, the fire in the bathroom did not die out but continued to smolder. Reason for the smoldering was that there was not sufficient oxygen in the bathroom to permit complete combustion and break the smoldering cabinet into flames and so it smoldered all night and into the next day—while the young physician lay unconscious.

In the incomplete combustion of a fuel—the smoldering state—large quantities of cabon monoxide are evolved. Carbon monoxide has no odor and will not burn, but it is fatal if inhaled in sufficient amounts. The sleeping Dr. Dee was subjected for many hours to a confined atmosphere heavy with carbon monoxide, and that is what brought about his death.

All the windows in the apartment had been closed—evenings can be chilly in Houston. When Dr. Jay and the custodian burst into the apartment the morning of the fire, that let a great blast of oxygen into the carbon-monoxide–filled rooms. The sudden union of oxygen and carbon monoxide in the right proportions causes an explosive reaction, liberating enough heat to cause the gas to flame, and that is exactly what did occur.

Everything learned during the investigation appeared to fit the above reconstruction of the happenings and their sequence. The insurance company attorneys took my report verbatim and laid it before the plaintiff's attorneys. They had a great wrangle over it for more than a week. The plaintiff brought in another fire expert during this time, but before we had a chance to depose him, the suit was called off insofar as the owners of the apartment complex were concerned.

I don't know what the final outcome of Dr. Dee's sister's legal actions was. At least in this case I was glad to absolve the role of aluminum house wiring in causing a loss of life.

The Case of the Boss's Romance

As World War II came to its decisive end in late 1945, there were many new technologies and devices developed and used by the struggling powers that were eagerly awaiting commercial development. As example of some of these that later came into wide use are: rocketry, which heralded space age explorations; jet engines, now used by all large aircraft; magnetic recording, first in the form of steel pin-wire as developed by the Germans and later as our well-known mylar resin tape recorders; and many others.

Of special interest in this connection was the development and use of radar during the seige of Great Britain. Television as such was still in the laboratory before the war. But the urgent need to see distant enemy objects regardless of weather spawned radar and solved the necessary, important elements of TV technology.

And so broadcast television came into being in our larger cities shortly after the war with test patterns and a few hours of programming—mostly old movies and how-to-do-its. To the average person the workings of TV were a complete mystery, and even the broadcasters were unable to cope with weak signal problems and the frequent multipath reception that produced picture "ghosts." TV antennae, the only way to receive any signal, sprouted on roofs and in many most unlikely places. The shadowing effect of buildings and large steel structures was not understood well and most often neglected.

At that time, the company I worked for was manufacturing wire and cable products. Foreseeing that demand for lead-in cable from rooftop antennae may build up into millions of feet, the company decided to open a new product line that could be very lucrative. Since I headed up the insulation research laboratory, we made a very thorough study of the propagation of TV signals through cable lines. Eventually, with the cooperation of RCA, we produced the earliest version of the flat wire TV lead-in line—familiarly known now as three-hundred-ohm line. The predictions for usage of this product were far surpassed by the huge demand for it—continuing to this day.

As a result of this development work and much study of television equipment research, I became fascinated by the new art and even built a TV receiver from scratch using an eight-inch green radar scope as a picture tube. This was the first unit in our town, and our living room was a very popular place for neighbors and friends when a program was on the air. And so I became the town's TV expert and of course also the resident TV expert in my company—and thereby hangs this tale.

One morning I received an unusual telephone call. It was from the big boss's secretary. She asked me to come up to his office when I found it convenient. That "when I found it convenient" seemed very odd. I called him "the big boss," but he was bigger than that. He was the president of the company, and I had never been called to his office before. At any rate, I hurried up to the executive suite and was shown into the president's office at once.

His greeting was quite cordial, and he asked me to sit down in the overstuffed leather armchair close to his desk. Then with a fetching smile on his face he said, "Gene, I would like you to do a personal favor for me, and what I want you to do is right in your field. It's not generally known," he continued, "that I have recently become engaged."

Here he hesitated, then said, "She is a wonderful woman,

everybody knows of her, and she is now producing a TV show for one of the New York stations. The station technicians have set up some receiving equipment in her apartment so she can become better acquainted with this medium that's new to her. But what they installed doesn't work well at all, and so far they don't seem able to improve the reception."

"I have told her," he continued, "about our own TV expert here, so I would like you to go down to the city, look in at her apartment, and see what can be done to clean up the trouble.

"Of course" he added, clearing his throat noisily, "call her and make a date when you have a free morning, but make it soon."

Back came the fetching smile and, "Will you do that, Gene?"

I told him I would be happy to see if there was anything I could do, but I said a downtown location might pose a real problem. I asked, "May I have her phone number and address. I will call her this morning."

"Good," he said, a broad smile now. "She is Gloria Swanson—how about that?" and I said, "Wow."

"When you are ready to go," he was back to business now, "call my secretary and she'll have a car and driver, who has been there often, at your disposal."

And that is how I came to know and admire at close range a world-famous film star. At that time Gloria Swanson had completed a long and very successful career as lead actress in many of the film industry's most popular productions. She had recently returned to New York City to a partial retirement, but soon became interested in TV and its possibilities. Of course she was deluged with requests to appear on the programs of the newly started broadcasters, but at the time I met her she had tentatively affiliated herself with one of the city's largest TV stations, perhaps financially as well as programmatically. This was the TV station that had set up the troubled receiving equipment at her apartment.

Miss Swanson lived in a ground floor suite of a very large apartment building off Fifth Avenue. A Filipino houseboy opened the door for me, and then Miss Swanson came forward and greeted me most cordially. She made me feel at ease quickly, asking many questions about my education, my work for the company, and my family. I could see that the apartment consisted of many charmingly furnished rooms. There were portraits of numerous movie personalities hung on the walls, many of whom I recognized. On the entire long wall of one room was a mural depicting the city of Paris. She told me that this was a gift from the people of Paris commemorating her starring in the film *Streets of Paris*. She talked most interestingly about various episodes in her career as an actress, and of course I was a rapt listener.

Finally we got around to the reason for my visit. She showed me the TV receiver, a monstrous cabinet affair with a little ten-inch picture tube and a great many control knobs. I asked about the antenna and found that it was located out in a courtyard, which was part of her suite. The antenna was mounted in what looked like a large planter box, and its maximum height above the courtyard paving was perhaps six feet. It was an elaborate dipole antenna with three director rods and as many reflectors, but there it sat in the courtyard surrounded by the towering walls of the building on three sides.

We went back into the TV room, and Miss Swanson turned on the receiver as she had been instructed. After three or four minutes the test pattern of the station came on with wavering lines, not as one picture but with repeated pictures offset to the right, one merging into the other. I counted at least three prominent ghosts and a number of lesser ones. Of course it was an impossible picture, and it was also clear that the problem was not in the TV receiver.

A TV ghost is a shadowy replica picture displaced a short distance to the right of the main or bright picture. It is caused by a part of the signal from the TV station being reflected from an obstruction in its path. The reflected signal path being longer than the direct path causes it to arrive at the receiving antenna a little later than the direct unreflected signal. This offset picture reduces the clarity of the signal and is very annoying. When there are more such ghost pictures due to multiple reflections from several walls, et cetera, the reception is almost unusable.

I asked, "Did the TV station people put up your antenna?"

She said she didn't believe so and thought that that part of the installation was made by a local radio shop.

"Well," I told her, "your reception problem is entirely due to the location of the antenna, and there is no way to overcome that as long as the antenna stays where it is. If we can place this same antenna, which of itself is very good, up on the roof of this building, your TV picture should be fine.

"The problem is," I continued, "almost all building managements in Manhattan are now forbidding placement of TV antennae on the building roofs. But I wonder, your being who you are, Gloria Swanson, that rule might be set aside. I suggest you contact the building's management agency and see what can be done."

At that, with a mischievous smile she said, "Oh, I don't think there will be any trouble with our management. Put up the antenna wherever you think it should go. You see, I own this building."

That floored me. I stared at her speechless and then couldn't resist breaking out in laughter—in which she joined, most heartily. And with that we went up the eighteen floors to the roof, where I was quickly able to select a clear line-of-sight location to her TV station, W——.

In partial defense of that radio shop's bonehead antenna installation I should repeat what was indicated previously, that with the advent of commercial TV broadcasting, antennae sprouted in great numbers on building roofs all over the city. These together with their lead-in lines, out of windows and running up the building walls, were not only most unsightly but also posed hazards of all kinds. The result was to ban all individual rooftop antennae on buildings. TV engineers were working on single master antennae for buildings, but that was still in the future. So it seems that the radio shop placed Miss Swanson's antenna in the only available open space they could find and left it at that.

The following day I returned to the apartment (in chauffeured luxury) armed with several hundred feet of our own lead-in cable, the necessary tools, and a plant electrician as assistant. Miss Swanson was very helpful in getting us up to the roof and planning how the lead-in line could descend to the ground floor. When the job was finished and we had assembled in front of the TV set, I turned it on. With the old vacuum tubes slowly getting up to transmission temperature, the picture was blank for more than a minute, but then it developed across the screen—a clean, clear, unghosted, fine test pattern. Success felt very good.

After that I had to stay for lunch and we had more good conversation. In particular she told me about her teenage son, who was having a troubled time trying to decide what to do with his life. He had some leanings towards a technical career, and she asked me if I would talk to the boy about the requirements and opportunities in engineering. I had a better idea and suggested, since it was early summer and his school was over, it might be good for him to spend four to six weeks working in my laboratory. His association with our research engineers could give him insights that he would not get any other way. Miss Swanson thought that was an excellent idea and said she would talk to my big boss about that possibility.

Before I left she gave me an eight-by-ten photo of herself as a young screen star and a pair of lovely Parisian earrings for my wife who has treasured them since.

With regard to young Swanson, it turned out as I had suggested. The big boss arranged the whole thing, and he spent a very interesting summer with us. As far as I know, he went on to college and enrolled in engineering.

Something happened to the big boss's engagement, and it appears to have been broken off. All I know is that the breakup had something to do with Miss Swanson's return to Hollywood shortly after I met her to star in the *Sunset Boulevard* production.

The Case of the House without a Ground

Washington is known as the Evergreen State because of its lush verdure, heavily forested mountains, great areas of matted wild rhododendrons and azalias, and countless lakes. But all of that is on the west side of the Cascade Mountains. This chain of high volcanoes extends from Mount Baker at the north end with Mount Ranier, St. Helens, and Adams marching to the south where they join the Oregon Cascades. The moisture-laden clouds moving in from the Pacific come against this Cascade barrier so that their fall of rain is largely between the mountains and the coast.

On the eastern side of the Cascades the annual rainfall is quite low, and much of this part of Washington is dry and desertlike. The natural growth is in bunch grass, tumbleweed, and sagebrush. Toward the north where the glacier tongues penetrated into the state there are stands of noble ponderosa pine, but these are closely restricted to the narrow gravel tongues. Of course eastern Washington is a heavy producer of wheat, alfalfa, apples, other fruit, and many vegetable crops, but with the exception of wheat, all must be under irrigation. Wheat and other grain crops are dry farmed, depending only on the meager rainfall and winter snows to carry them through to bumper yields.

Those counties that lie just east of the Columbia River are particularly arid. Their good fruit, alfalfa, and sugar beet production is based on an extremely large irrigation system requiring heavy pumping capacity and consequently heavy power availability. The public utility districts that supply electric power in these counties operate several large hydroelectric plants on the Columbia River and also purchase additional power from the federal generating plants in Washington.

Except for the irrigated fields and a few industries, the small townships and the generally sparsely populated land areas require comparatively little electric power. However, county residents are spread very far apart and this is reflected in the very long distribution lines necessary to supply each house. This of course increases the cost per kilowatt of energy utilized.

There is another problem in distribution. Electric supply systems can generally depend on a ground return to complete the system's circuit. The earth (ground) represents an almost infinite conductor of very low resistance so that *well-made ground returns* at various service points tie them together better than if the returns were made by means of large copper conductors. However, in an area where the ground is very dry and dry to an appreciable depth it may not be possible to establish a well-made "ground." If this condition exists along a distribution line, it is necessary to include not only the energy bearing phase conductors, but also a so-called neutral wire, which forms a positive ground return for the various loads served along the way. The neutral wire of course is tied into the solid ground of the supplying generating or substation. Most utilities provide a neutral conductor on their distribution lines without regard to earth conditions. Providing such a neutral adds only moderately to the line cost.

Getting back to the dry counties of Washington and specifically to County X, the PUD there had to provide service to various outlying sections, some of which were ten miles or more from a substation. Also, there could be perhaps only two or three residences at the end of one of these long runs. In addition, some of the services were for mobile or motor homes that were likely to be moved from time to time. Under the above circumstances, the

PUD elected to put as little money as possible into these extended, lightly loaded services and so they ran the pole lines out with only one energy-bearing phase conductor and *no neutral wire*.

One hot summer day in July a few years ago I had a call from an attorney representing the PUD. He spoke about a fire that had completely destroyed a residence served by the PUD, and he wanted me to look into any possibility that might involve the PUD.

The circumstances of the fire appeared to be very unusual. On the day of the fire there had been a heavy westerly windstorm clocking up to sixty miles per hour and blowing hard for almost ten hours. Damage was generally not serious, but it was widespread. However, in one area the brunt of the savage wind was at right angles to a section of one of the long, twenty-five-thousand-volt single-phase pole line runs and it eventually brought down the line conductor of a span happening to be in front of a residence. The conductor, a No. 6 hard drawn solid copper wire, parted near the pole on which was mounted the step-down transformer supplying that house. The break was so close to the transformer pole that the conductor was too short to reach the ground and it just hung down banging against the pole in the wind gusts, about five feet clear of the ground. The other end of the broken conductor of course lay on the ground, but being on the load side of the service, it was unenergized and caused no trouble.

There was *no* neutral wire on this line. In order to complete the circuit for the residence's electric supply the PUD had provided a bare grounding wire running from the transformer down the pole to a so-called butt ground. This type of grounding is made by passing a number of turns of the bare wire around the bottom portion of the pole that is buried, usually about six feet. In normal earth this type of grounding can be passable, but in a desertlike situation it is virtually worthless.

Thus the power supply to the residence consisted of the live secondary low voltage taps from the step-down transformer, plus a bare wire extended from the butt grounded wire on the pole. The residence had been served in this manner only a few months.

At the time that the line conductor break occurred, the occupants of the house, husband and wife, were at home. Apparently they were horrified to see and hear sparking and arcing at every wall plate switch, at the lighting fixtures, which promptly blew out, and at the sink and tub fixtures. They rushed out of the house and indeed were very fortunate to be able to do so. With a great explosive sound, the house literally burst into flames and was totally destroyed in a matter of fifteen minutes or so. From a neighbor's house, a quarter of a mile back toward town, calls were made to the fire department and to the PUD. And so about ten minutes from the time the conductor break had occurred, the line was finally killed. The fire truck didn't arrive until a half hour later.

The situation when I got involved for the PUD was that the owners of the destroyed residence had entered suit against the PUD, claiming not only a full replacement cost for their home but additional punitive damages for shock due to extreme negligence. As it turned out, I was not involved very long in this case, as the reader will see shortly.

After spending about half a day at the PUD offices talking with their several engineers and construction foremen and going over their distribution line drawings, I went out to the site of the fire. By that time I had a pretty good idea of what had transpired that windy day when the line came down. Of course by that time the downed line had been repaired and resagged and except for burned house debris where the residence had been, everything was as it was before the line break.

I had with me a ground resistance measuring instrument and three four-foot copper grounding rods. These we drove into the ground around the distribution transformer pole in a prescribed configuration. Applying the well-known three-electrode method, I determined the ground resistance at two locations. I expected it to be high, but it was a good bit higher than what I had in mind. It was surprisingly high, so high, in fact, that driven grounds, pole butt grounds, or even elaborate counterpoise ground systems would not provide a reliable return circuit.

Knowing this, it was simple to reconstruct what happened out there when the live conductor in that span parted and hung down along the pole. It kept banging against the bare grounding wire affixed to the pole between the transformer and the pole's butt ground. At any time that it touched the ground wire this should have caused a grounding relay back at the substation to trip the circuit breaker there and deenergize the line in a matter of several tenths of a second. But the grounding relay could not operate because the current to activate it could not flow back in that excessively high resistance earth.

Therefore, each time wind gusts caused the broken line to contact the grounding wire, it sent a pulse of high voltage, up to 14,400 volts, into the house. Since the house electrical system insulation was designed (by Code) to sustain a maximum of only six hundred volts, the high voltage pulses broke down the insulation at every switch and electrical fixture and the insulation along the house wiring and also energized the entire plumbing system. The sparking and arcing must have ignited the walls and ceilings in seconds, and in a short time the whole house was burning.

When we got back to the PUD office, the manager and his staff listened rather grimly to my explanation of the fire cause. Of course it placed the responsibility for what happened squarely on the PUD. I recommended that no distribution lines run in this high ground resistance area should be without an adequate-sized neutral wire and that the ground-fault circuit breaker relays be set and tested to be responsive to faults occurring at the end of any such line.

I could see at the end of our conference that I had become persona non grata. They thanked me rather stiffly and asked me to send in a bill for my services to date.

In the weeks that followed I heard nothing from the PUD or from their attorney. Finally I called the attorney and told him I had written a report of my findings and asked if he wanted a copy. He replied that he did not want the report but I was to hold it and any file I had for possible later use. After several years had passed with no communication whatever from my client, I closed the file on it. Evidently this consultant did not have the desired answers.

The Case of the Double-set Poles

In rural areas between cities and towns, treated wood poles still carry the bulk of the electric power required. These are fairly tall, forty to seventy-five feet high, and on them are strung power lines at the so-called subtransmission voltage levels of 35,000, 69,000, or even 110,000 volts. Such poles are generally placed along the roadways on prescribed easements a few feet away from the roadside ditches. These pole lines are a common sight along our highways, marching up and down hills, crossing over intersections and streams, and running for many miles to substations where they terminate—and restart in a new direction.

They are multiuse poles. The high voltage power lines are carried on crossarms near the top. Considerably lower can be telephone and telegraph lines, and more recently cable television circuits also have been included.

Wood poles set in the ground must be treated to prevent damage to the butt caused by earth bacteriae, mold, insects, ground water rot, et cetera. Depending on the soil's nature, an untreated fir pole set in the ground could become unusable and in danger of breaking off after ten to twelve years of service.

When the butt of the pole is treated with suitable chemicals under pressure, its life can be extended by ten or twenty years. However, any pole in service for more than twenty years is suspect.

All power companies have numbered metal plaques nailed to the pole by which a cruising lineman can readily determine installation month and year. If he spots an older pole, he digs around its base a bit, looking for rot or fungus attack. Should he find sufficient damage as to possibly render the pole unsafe, he will report that pole for replacement.

Replacing a pole in service carrying energized high voltage conductors is not simple. The new pole must be set in almost against the old pole but in such position as not to interfere with crossarms and wires. Then the task of transferring all the conductors to their proper places on the new pole begins. First the required number of new crossarms and insulators are attached to the new pole. Then one by one, in a proper order, each conductor is moved over and clamped. Rubber insulating gloves, insulating blankets, and hot-sticks must be used expertly in this operation, since the transfer usually is made with the wires still energized. Power is shut off only if physical conditions and pole configurations require splicing in additional conductor lengths or some other unusual situation exists.

In view of the personnel and time demands for replacing a service pole and where the butt of the old pole is not too badly damaged a quick means to reenforce it is often resorted to. By this method a treated stub-pole, so-called because it is only ten to twelve feet long, is set in flush along the old pole and a number of steel bands are wrapped about both poles and tightened in with special tensioning tools. This scheme works fairly well and gives the company time to schedule a full replacement when most opportune.

With this short preamble on power poles we are ready to go ahead with my story of the double-set poles. (If the reader wants more information on power poles, their specification requirements, installation, et cetera, he should look at a current edition of the National Electric Safety Code.)

The accident happened in a rural area near a small city on the northwest coast. It was early evening in the summer, and visibility was good. Two cars were approaching an intersection.

The car going west was on a main east-west highway, and the young woman driver was proceeding at perhaps fifty-five to sixty miles per hour as she approached the intersection. The second car, with two college students on vacation, was on the smaller north–south side road and approaching the stop sign at the intersection from the north with the intention apparently of turning to the right onto the westbound highway. Whether they actually came to a stop at the painted stop-bar and then turned will never be learned. We do know that the speeding westbound vehicle slammed into the college students' turning car with tremendous force and both cars ended up in the ditch on the south side of the highway about fifty feet from the intersection.

When the rescue squad arrived they couldn't separate the two cars. That had to be done later with a cutting torch. The human toll was: the college student driving the turning car died in the wreckage; his passenger was alive but very badly injured; the young woman in the speeding car was injured but not nearly to the extent of the others.

Later it was learned that the woman had had several cocktails during the afternoon before the accident. She admitted to having exceeded the forty-five-mile-per-hour speed limit posted on that section of the highway but insisted that she had the right-of-way. The badly injured passenger in the other car was a long time under hospital treatment and a long time recovering some use of his limbs. It was almost six months after the accident before he could be interviewed about what had happened. He was very hazy as to whether they had come to a stop or not before turning at the intersection. He did say they did not see any car approaching the intersection on the highway and thought it safe to make the turn. He was pretty sure of this throughout the questioning.

The families of the two young men engaged legal counsel to determine what redress they might have against those directly and/or indirectly responsible for the accident. Their attorneys visited the scene several times and took photos of the intersection and the four approaching roads. They then issued a statement that the college students actually could not have seen the approaching car on the highway because there were two tall power poles side by side on the northeast corner of the intersection, which effectively obscured the highway vehicle. On this basis, suit was entered against the municipal power company, the county roads commission, and the woman driver. The complaint against the county roads commission was that they permitted the placement of sight obstructions at corners of road intersections. The complaint against the municipal power utility was that they had set up two power poles close together at the intersection corner that made it impossible to see an approaching vehicle on the other road. The suit against the woman driver was perfunctory. They didn't hope to get much from her.

That was the situation when the utility called on me to assist their engineering staff in getting at the facts. What they needed from me was to develop a nationwide background on the setting of power poles at street and road intersections. Also, I would look into the double pole/obstruction occurrence at that specific intersection of the accident.

For the survey of corner pole setting I prepared a questionnaire that was sent out to some fifty large and small power utilities. Basic to the inquiry was a request to state their practices in pole line construction.

Getting into the double pole picture, I went over the utility's line construction crew records and interviewed the principals involved in their line installation and maintenance work. I learned that most of the poles of the transmission line running along the north side of the highway were some twenty-five years in service. Consequently a number of these had been marked for replacement, including the pole at the involved corner. About three weeks before the accident, the line crew had set in a full-sized new pole against the old pole on that corner. It was anticipated that they would come back in a day or two to transfer all the crossarms and conductors to the new pole and pull out the old one. But there were delays and then extended delays so that on the day of the accident the two poles were still in place.

Why these delays? I found the answers. That pole line not only carried the utility's power conductors but also telephone and cable television lines. The utility had notified each of the other companies on the date proposed for pole replacement and had received assurances of full cooperation. But first the telephone company and then the cable TV company started to set new dates because of unforeseen commitments, they said, and so nothing was done until shortly after the accident, when the old pole was completely replaced.

Could two side-by-side ten-to-twelve-inch-diameter round poles be located at such a critical corner position as to truly prevent a car driver moving toward the intersection's stop sign from seeing another vehicle approaching on the crossroad? I spent hours at that intersection at various times of the day, moving back and forth on the west side of the side road and looking and photographing oncoming cars on the highway, always keeping the pole in my line of sight. In no way could that pole obstruct my sight of the highway vehicles. Then I asked the utility to remount a stub-pole, one long enough to stand at least ten feet above ground alongside the corner pole to simulate their widest possible condition at the time of the accident. Again I moved up and down the side road, keeping the double poles in line of sight and checking if oncoming vehicles on the highway were concealed. Again I could find no position near the intersection at which the double pole effectively obstructed sight of highway vehicles.

There was good reason for my findings. The distance between the corner pole location and a southbound car on the side road approaching the intersection was approximately thirty feet. At that distance with normal binocular vision there could never be a time for the poles to block sight of highway vehicles. Even with monocular vision the oncoming vehicle could only be partially obscured and that only for a fraction of a second. However, from college records we learned that the fatally injured driver had excellent binocular vision. We also learned that he lived only a short distance from the intersection of the accident and at the time was taking a college friend to his home for a visit. Therefore, he knew the area pretty well and was certainly aware of the side road stop requirement at that intersection.

As a result of my investigations I came to the conclusion that the two college friends were chatting, laughing, or whatever as they drove toward a familiar intersection, never looking to see what was approaching on the highway, and turned onto it disregarding the stop sign and the stop bar on the pavement. Had they been alert, paying attention, they would most certainly have seen the speeding car on the highway.

I reported my findings and conclusions to the utility's attorneys. Also the answers to our questionnaire sent those many power companies were coming in steadily. With no exceptions they reported on the common practice of placing poles on prescribed easements near street and road intersections. Also, no accidents resulting from a pole visually obstructing other vehicles on adjacent or intersecting roadways were reported. Corner placed poles were statistically somewhat more frequently involved in being struck by cars than others, but no change in practice was being made on this account. Thus the local municipal power utility was in conformance with nationwide practices on pole placements.

A number of interrogatories were exchanged between the attorneys and several depositions taken, including mine twice. Then it appeared that the plaintiffs were interested in a quick settlement of their claims. However, both the utility and the county road commission decided to let the matter go to trial. And it did.

Of course I was the expert witness for the utility and gave my testimony as I have indicated above. The trial was over in three days, and the jury deliberation took only a few hours. The defendants, municipal power utility and county roads commission, were held blameless. The woman driver of the highway car was found guilty of excessive speed and careless driving.

Moral: Stop, look, and listen.

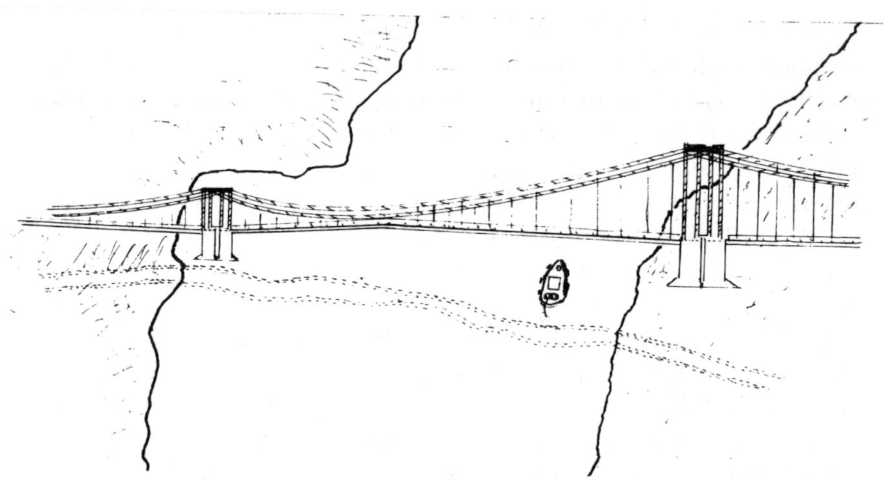

The Case of the Ancient and Honorable

There is no rationale for the capricious inconsistency in the value of old things. For example, if you go into a thrift shop (new name for secondhand store) you can probably purchase a set of very decent-looking breakfast plates for twenty-five cents apiece or a four place settings of stainless tableware, not too worn, for a couple of dollars. But if instead of a thrift shop you were in a "House of Antiques," those plates, a bit more stained and chipped, might cost $2.50 apiece and the tableware could be quite unaffordable.

The situation is reversed when dealing with high technology apparatus of any kind. Particularly in the field of electronics, a device can become obsolete in a matter of one year or even less. And once it is replaced by a fancier, more automated, do-more unit, you can hardly give the "old" one away. However, there is an important exception to such rapid depreciation, one that is not too well understood by people who should know it. Basically the electric cable buried in the ground or in a conduit or underwater in a river or lying on the bed of an ocean so long as it continues to operate satisfactorily has *no* value depreciation. The case that follows is about an electric cable, in service for many years, that was wrongfully tagged as obsolete.

The story goes back to the year 1939, when a west coast power company replaced a very long roundabout transmission pole line serving a coastal community with several such submarine cables across a two-mile-wide canal. These were top-line rubber insulated cables with steel wire armoring made to state-of-art standards of the time.

When laying submarine cables in navigable waters the state, the Coast Guard, the Coast and Geodetic Survey, and more recently the Environmental Protection Agency all must have details of the laying plan and the exact locations. If approved, permits will be issued that spell out the responsibilities of both the power company and the state.

All this was done in 1939, permits issued and the cables laid and placed into three-phase, fifteen-thousand-volt operation. Maps showing the cable crossing remained on file with the state and the Coast Guard as required by statute.

With practically no service interruptions these cables remained energized continuously until March of 1982. At that time they became irreparably damaged when fouled by several bridge anchors and had to be abandoned. The circumstances resulting in this premature loss of serviceability were as follows:

In 1978 the state had authorized the construction of a vehicular bridge across the canal and awarded the contract to a large, well-established highway and bridge construction company, which I will designate as the Bridge Company. As required by the terms of the contract, the Bridge Company investigated every aspect of the crossing shorelines, water flow, and bottom conditions, obtaining this information mostly from records submitted by the state and the federal coast agencies. Also stated in the contract, the presence of certain live submarine power cables in the crossing area was called to the attention of the Bridge Company—to exercise special care in operations close to where they lay. Further it was stated that should these cables be damaged by the Bridge Company, they would be liable for all costs for repair or replacement as verified by Power Company owners.

Power Company provided Bridge Company with details of these cables and their locations as originally laid. Although this was not documented, they also advised Bridge Company that due to bottom currents the cables could have shifted from their original positions a small but undetermined amount.

Everything went ahead smoothly until the canal bridge was almost finished. In early 1982 Bridge Company began pulling up their heavy construction anchors and buoys using tugs with powerful take-up winches. During the week this activity was going on full blast, one morning a failure siren sounded in the utility's substation from which the submarine cables were energized. With the alarm, the circuit breakers feeding the cables opened up, indicating a massive short circuit somewhere. It was quickly determined where the failure was located. Opening the far end of the cables and checking for their insulation condition immediately showed both cables to be shorted to the canal waters. In other words, they both had been badly damaged. Using a radar fault-locating device, the damage area came out almost exactly where a Bridge Company tug was seen to be hauling up anchors at the same time the substation alarm went off.

About a week after this happened, Power Company called and asked me if I could determine the condition of these old cables at the time of the service interruption. I suggested that they cut out a twenty-foot section of each cable from their underwater portions and that we make a dissection of these pieces. This was done and I carried out a careful visual examination of the various cable components. What I found was quite surprising. These cables had been made in 1938 and had been in constant service for nearly forty-two years. Except for some minor rusting of the steel wire armor, the cables showed practically no signs of any deterioration. In particular, the oil-based rubber insulation appeared firm and free of any electrical discharge paths or the slightest amount of carbonization.

On the basis of this finding, Power Company decided to bring a law suit against Bridge Company for replacement of the submarine cables and other costs. Previous discussions with them had been very unsatisfactory. Bridge Company disclaimed responsibility for damaging the cables and also said that the cables had already outlived their useful life and had very little remaining value.

In view of the remarkable condition of the cables, as represented by the sample lengths I examined, and the pending law suit, I suggested that additional lengths of the cables be cut out and sent to a commercial cable testing laboratory and there be subjected to the same physical and high voltage tests as required for new cable prior to shipment from its factory. I was gambling a bit on this, but I had a deep feeling these old cables could still pass the same factory tests they had been subjected to forty-two years ago.

Well, we did send cable samples to a cable-testing laboratory and they were tested in accordance with industry-specified conditions for that type of new cable. They passed all these tests with a comfortable margin, and we received a certified test report from the commercial laboratory covering their work. I was delighted with the test results because I could now state with reasonable certainty that at the time the cable damage occurred they were in excellent condition, with a life expectancy equivalent to that of new cable.

In the following months the legal maneuvers of the two parties ground on. After Power Company entered its complaint and claim for damages and Bridge Company its counter claims, there were many interrogatories issued by each party and depositions taken of experts and company personnel. In my two depositions I said I had no knowledge of the circumstances that caused the cables to be damaged. Instead I concentrated on my examination of the cables, their excellent condition, and their probability for giving many more years of normal service had they not been damaged severely. Also I gave some statistics on cables, installed worldwide, that had given satisfactory service for thirty, forty, fifty, or sixty years and even longer. Most of these were lead-sheathed impregnated-paper cable types, but a few were rubber-insulated.

I was not asked nor did I volunteer to provide some rather startling new information I had uncovered a few weeks before. During my earlier dissection of the two cable samples, the thickness of the insulations was measured as a matter of course. Comparing this value with industry specifications verified that they were rated for twenty-five-thousand-volt operation. Beyond noting this, I paid it no further attention. However, in later discussion with Power Company's engineers they confirmed that these submarine cables had been in fifteen-thousand-volt operation throughout their service lives. I asked why they had purchased twenty-five-thousand-volt cable to operate in a fifteen-thousand-volt circuit. Oh, no, they said, these cables were rated for fifteen-thousand-volt service and were so purchased. I asked to see the purchase requisition, but it was so long ago that it couldn't be found. Well, when I showed them how the insulation thickness was related to voltage rating and in this case the thickness was much greater than required for fifteen-thousand-volt operation, they reluctantly agreed I was right. So it was that these twenty-five-thousand-volt cables had been operating at only 60 percent of their voltage capability for forty-two years. Operating at reduced voltage is a significant factor in extending cable life, and I went into this later at the trial.

I should mention that our attorneys had deposed a number of Bridge Company's employees and their experts and I was able to study the transcripts of these depositions. They had brought in four experts to assist their case. Three of these were construction/mechanical engineers with great knowledge of bridge construction and anchor- and buoy-setting procedures. The fourth was their cable expert. But he was a retired communications engineer with experience only in telephone and telegraph systems and admitted he had no experience in the power cable field. I could see he would be of little use to his client.

As the trial opened in a superior court, the contentions of the two parties simply said were as follows:

The Power Company:

1. The cables were irreparably damaged by Bridge Company.
2. Bridge Company was apprised of the cables' location but warned that they may have shifted due to bottom water currents.
3. When damaged the cables were in excellent condition, with many years of service life remaining.

The Bridge Company:

1. We didn't do it.
2. Those cables had long outlived their useful life and were ready for retirement.

I was on the witness stand for only about two hours. On direct examination I described in popular terms what a power cable was, how it was manufactured, what determined its service rating, and how it was modified to serve in submarine operation. Then I told of my finding the cables to be in virtually unused condition and confirmed this by having them successfully tested as though they were newly manufactured. From this I unequivocally stated that these cables could have continued satisfactorily for many indeterminate years. Besides citing industry records for cables that exhibited exceptionally long operating service without failure, I threw in the bombshell that the damaged cables had never been subjected to the voltage for which they were designed, but were operated at a much reduced voltage level. I then explained that insulation deterioration was very sensitive to voltage stress level and for rubber-insulated cable subjected to one-half its design stress it is estimated that its life would be increased by a factor of four to six times—in accordance with a power law relating voltage to life.

I believe that did it. The jury found that Bridge Company had damaged Power Company's cables and that at the time they were in new condition. They awarded Power Company a reasonable amount to replace the canal cables. They did, however, reduce the total amount of Power Company's claims because of the uncertainty of just where the cables were lying at the time of the bridge construction.

The Case of the Crossed Connections

Setting up a major new submarine base is a vast undertaking by our government, and when it is a nuclear submarine base the facility complexities increase fivefold.

Construction was started on such a base some ten years ago at a west coast location. While the navy prepared all the designs and area plans for ship and munitions handling and in the main supervised its own construction crews for their installation, there were many other necessary base requirements that were let to private contracting companies. For example, the electric power supply, housing for about fifteen hundred base personnel, and electricity distribution to these, base roadways, recreation areas, landscaping, water and sewage facilities, et cetera. What I wish to relate here has to do with one of the residential areas where, as it turned out, an almost impossible situation existed.

Let me describe these residential areas to clarify the ensuing events. All of the civilian base workers, as I recall, lived in houses contained along six or seven deep courts. Each court was in a U shape, with dwellings facing from both sides onto a wide roadway forming a U. Access to these courts was by a good-sized highway running along the open ends of all of them, then terminating in the county's main road into the base.

Besides the some twenty houses, each court had a convenience store and a multipurpose clubhouse, and at the time of which I am writing, construction was still continuing, with only three courts completed and occupied.

Power to the base electrical substation came in on a three-phase, 115 KV high line. At the substation, transformers reduced the voltage to twenty-five kilovolts and this formed the primary three-phase distribution system for the whole base. Distribution of the primary was in the form of underground cables throughout. In the courts, each group of two or three residences was supplied through a closeby surface mounted transformer that converted the 25-kilovolt three-phase primary to a 120/240-volt single-phase secondary. The convenience exchange and the club house required higher power levels and were supplied by three-phase at secondary-use voltages.

Now just a quick word about electric power distribution systems and how they are arranged to provide a maximum of continuous service, free of interruptions. In the so-called radial distribution system, a primary supply line goes directly alongside all the loads to which it is to provide power and at each load the line is tapped off through a step-down transformer to that load. At the far end the primary is then terminated. It is a simple system using the least amount of line wiring. However, it has an important disadvantage. If the primary line sustains a fault somewhere along its run, the substation breaker opens and until the fault is cleared none of the loads will be served. Even after the downed section of an overhead line or the faulted portion of an underground cable is cut out, only the loads between the line damage and the substation can be served. Those beyond the break must await completion of repairs, and in the case of an underground cable this could take some time.

A better distribution method is known as the loop system. By this method of serving a given set of loads, the primary from the substation forms a loop around the loads and comes back to its starting point. In this way any load along the line can be supplied from either or both of two primary sources. Should a fault be sustained somewhere along the primary line, once it is lifted or cut out, the loads to either side of the fault position can be served from the loop legs until repair work is initiated. Loop distribution is more costly but provides better service continuity.

The residence courts at the submarine base were each served by looped primary cables. The large primary feeder cable ran in ducts along the highway fronting the courts. At each court the feeder was tapped by cable running in a loop around the court and coming back again tapped into the feeder. To augment the loop system a sectionalizing switch was added between the feeder and the looped cables at each court. The purpose of this switch is, if a primary cable failure should occur, to facilitate removing the damaged length while continuing to supply power to the longest possible section of the loop loads.

For this voltage the sectionalizing switch was fairly large and housed in a lockable steel cubicle at the front of each court. There are actually two fused switches in the assembly, each a three-phase (three-bladed) switch with the line or "hot" jaws of both switches connected to the primary feeder. The movable blades of one of the switches are connected to the looped primary on the north side of the court. The other switch blades go to the looped primary coming back from the south side of the court. Normally (and this is important in the light of the accident) power to the loop is supplied with only one switch closed, since closing both switches, although possible, serves no purpose. If the loop is being supplied by the north section switch (the south section switch open) and a primary cable fault occurs somewhere on the north side of the court, a fuse or fuses in that switch will open, leaving the court without power. The base electrician will be notified immediately; he will open the north switch, then locate the failure, cut it out, and then close in the south section switch. In this way a maximum number of court loads can be served at the same time that repairs to the faulty section are being made. So much for background.

My involvement in the submarine base came when an attorney representing one of the country's largest electrical contractors called me to assist in determining cause of an electrical accident at the base. He was very vague about the details of what happened but said that an electrician had been badly burned and there was some kind of involvement with a sectionalizing switch. I did meet with the attorney and learned that the accident had occurred some seven months ago. A primary cable on the north side of one of the completely occupied residence courts had failed. The base electrician foreman and two helpers had come down quickly to the court to restore power. They had isolated the fault and cut it out and then, following standard procedure, closed in the south section switch to provide power up to the cable disruption. Following this, they obtained an appropriate length of spare primary cable from stores and spliced it onto the dead end of the loop. They then opened the south section switch, completed the splicing of the spare cable to make the primary loop continuous, and again closed in the south section switch. This restored service to the court, and all seemed well at that point.

From here on recital of events appears confusing. The day after service had been restored, the foreman and an electrician helper came back to replace the blown fuses in the north section switch. These fuses are large, heavy cylinders sized to carry up to 150 percent of the maximum primary single-phase current loading of the court. With his key the foreman opened the cubicle door, exposing both sectionalizing switches. There was no interlock relay system on this door, so opening it had no effect on the energized switches contained in the cubicle. It is to be remembered that on such a 25,000 volt system each phase or line has some 14,800 volts to ground and 25,000 volts between phases or lines and these were the voltages across the switch parts.

Looking at the heavy fuses and the large, very sturdy clips they were to be forced into and knowing he would have to work close to high voltage components, the foreman decided to have the whole twenty-five-kilovolt feeder deenergized before putting in the fuses. Of course this would shut down power for all of the residential courts, but he felt his safety required that action. So he sent his helper up to the main base substation to have the feeder circuit opened. Both the foreman and his helper had walkie-talkies, and the helper was to let the foreman know when the feeder was deenergized and stay at the substation to receive the foreman's later instruction to reenergize the feeder.

Therefore, when the fuses were being inserted the foreman was alone. He got them in place properly, it seems, and, stepping back, called to his helper to reenergize the feeder. This done and acknowledged, he was about to close the cubicle door when he remembered that the court had been fed from the north section switch before the cable failure and was now being fed by the south section switch. (Apparently there was some arbitrary decision made to supply all the courts normally from the northerly switch sections.) So he decided to make the change over in the simplest way he could think of without disrupting power to the court. That was by closing in the north section switch, both sides of the loop now being connected to the feeder, and then opening the south section switch. Still alone, he proceeded to do this. The instant the north section switch blades touched the jaws there was a tremendous explosion and heavy arcing, which melted up the switch and part of the steel cubicle.

Fortunately, although the foreman was badly burned, the explosion had thrown him clear of the arcing and he was still alive when his helper found him twenty minutes later.

All that was seven months ago. Now the foreman, after spending many months in the hospital, was back in his home, incapacitated but alive and reasonably comfortable. He had just entered suit for recovery of his injury sufferance and lifelong impairment. Defendants cited were the several electrical contractors involved in design and installation of the court's electric system, the manufacturer of the sectionalizing switch, the manufacturer of the court's primary cables, and the power company supplying electric power to the base.

With respect to this area of the base, the electrical contractor who called me was responsible for the design, layout plan, and material inventory for the court. However, a local contractor had made the actual installations. Many depositions had already been taken—of the base electrician helpers, the foreman, the construction crew of the local contractor, and the engineers responsible for design and layout plans. So I had plenty of reading to do before even visiting the site of the accident.

Finally I was cleared into the submarine base and looked over the residence courts and the specific one involved. By now a new sectionalizing switch had been installed and all appeared in satisfactory working order. The new electrical foreman opened the new switch's cubicle, and I saw that again the court was being fed by the north section switch with the south section switch open. I asked the foreman if we could have the main primary feeder opened at the substation for a few minutes so I could use my resistance meter to check the phasing on both switches. He asked me why I wanted to do that. I replied that I wanted to see if each of the three phases on one switch was connected to its corresponding phase on the other switch. He thought about that a bit, then broke into a grin. "I get you," he said. "If the phases don't match there would be a h—— of a short circuit when both switches are closed in. Well," he continued, "a little while after this new sectionalizing switch was put in I had occasion to close in both switches. Nothing happened and I didn't give it any thought. Say, I'll close them both in now if you wish."

"Wait a moment," I said. "Are you absolutely sure that everything was energized when you closed in both switches before?"

"You bet," he said.

"One more question," I put to him. "When this new sectionalizing switch was installed, were all the connections to it remade?"

"Yep," he said. "Remade and checked by the contractor's engineer."

"Okay" I said, moving back an appreciable distance. "Go ahead and close in the other switch!"

He did; nothing happened. He did it several times; there was not even a spark as the blades met the jaws. That certainly answered my question and, further, reenforced a growing suspicion of how the accident had occurred. I reread the statements made by the cable-installing crew and the statements and planning prints made by my client's staff. Then I asked our attorney to find out from the injured electrical foreman if the two switches in the sectionalizer had ever been closed at the same time prior to the time of the accident. The answer came back unequivocally: they had not! That residence court had been completed only a short while before the accident occurred, and there had been no need to change the serving points for the primary loop up to the time they had the cable failure.

Now I was convinced that the original sectionalizing unit had in some way been misconnected. The three-phase conductors to the two switches had not matched. Phase A on the north section switch was connected to phase C on the south section switch, and its phase A was connected to phase C on the other switch. Furthermore, these misconnections were made on the movable blade side of the switches that is on the side connected to cables going around the court loop. Therefore, as long as only one of the sectionalizer switches was closed there was no problem, but closing both switches together would cause an instant dead short circuit between phases A and C—a short circuit across a twenty-five-thousand-volt circuit. No wonder the whole switch blew up.

I gave my reconstruction of the accident at a meeting with our attorney and several other expert witnesses who were assisting our client. There was little problem in convincing all present how and why the accident had occurred. All agreed that an unprecedented misconnection had been made and in some way both electrical contractors were involved. This was communicated to the plaintiff's attorney, and they promptly eliminated the power supply company and the manufacturers of the sectionalizing switch and the primary cable from the suit.

I never did learn how responsibility for the actual incorrect wiring was fixed—or if it was ever fixed. The case never came to trial. A negotiated settlement was made by which both electrical contractors shared, in some way, the settlement costs.

Perhaps this was a reasonable solution. Although my client's design and plans for the loop circuit were correct, and I had verified this by examination of the pertinent documents and drawings, he did have the responsibility to inspect and monitor the construction and installation work of the local contractor. Evidently that responsibility was not fully carried out.

The Case of the Judge Who Said No

After retirement in 1973, my wife and I often spoke of travel to the West Indies, but we were on the west coast then and it was a long and expensive journey. Nevertheless, during the following years we did get to Puerto Rico, but not as tourists on vacation. I was employed as a consultant by a mainland manufacturer and had a job to do there. However, it was an enjoyable experience, the island setting was beautiful and fascinating, our beach hotel was sumptuous, and best of all, my wife was with me. Also, what I had to do turned out to be successful.

The one-thousand-mile flight from Miami over the bluest of oceans dotted with brilliantly white little islands and keys is one I remember with pleasure. At the busy San Juan airport you are lulled into thinking this is an American island, just like Hawaii, and everyone speaks English. Once you are out of the airport, no one on the street seems to speak or understand English and you realize you should have brought a Spanish dictionary along. Of course Puerto Rico is American, but for four hundred years it was under Spanish rule and it wasn't until 1917, twenty years after the U.S. takeover, that Puerto Rico had a civil government and public schools. Very obviously the three or so generations that followed have had small effect on the language of the masses.

The problem that brought me to Puerto Rico was an unusual one. The Puerto Rican power authority had undertaken to provide electric service to a small island lying some twelve miles east of the Puerto Rican mainland. This island had become a popular residential retreat, with beautiful beach vistas and a quiet, peaceful atmosphere. However, its local oil-fired generating facility was unable to keep up with the increasing demand of new consumers. Their petition to have mainland power supplied to the island by submarine cable had been approved by the territorial government. To carry this out the power authority with the help of its consultants designed an underwater transmission system for operation at thirty-three thousand volts, three-phase. The plan was to carry the transmission line overhead on poles to the takeoff point on the Puerto Rican shoreline, then lay the twelve-mile submarine cable, bringing it up to the island, where it would be terminated at a distribution transformer.

Purchase specifications for the submarine cable were sent to a number of suppliers, with the requirement that the successful bidder not only furnish the cable but also carry out the laying operations. The AB Company of New England, a long-time manufacturer of submarine cable, was low bidder and accepted for the job.

The selected cable design called for three suitably insulated conductors to be twisted together to form a three-conductor cable, over which were several layers of filled, strong fabric tape as a bedding for the overall helically laid, thick steel armor wires. This cable was a so-called wet design, since it was intended that sea water would freely surround and contact the insulations. However, the insulation used was EPR (ethylene propylene rubber) a synthetic polymer known to have very little moisture absorption.

At the time the Puerto Rico order was placed, the AB Company was not in a position to supply the individual insulated conductor wires. It subcontracted the manufacture of these with another cable company. When the cores were made, tested, and shipped back to the AB Company, it completed the operations of twisting the cores together, applying the bedding tapes and the armor overall. Also, the AB Company carried out the operation of joining the as-made cable lengths together to form the twelve-mile required submarine length.

Since it figures importantly in the problem I was asked to look into, a word about cable jointing, or splicing, as it is more usually termed, is necessary. A cable must continuously connect two or more separated regions in space regardless of how far apart these regions are. There are, however, limitations in manufacturing long lengths imposed by the amount of material a given type of cable requires. The power-cable–making machines can produce long, continuous lengths of small size cable, but this is drastically reduced as cable size increases. For the submarine cable being made for Puerto Rico the longest finished lengths were in the order of four thousand feet. This meant the twelve-mile underwater link between Puerto Rico and the island would have sixteen full cable splices. And since it is much safer to make splices under factory conditions rather than in the field, the AB Company made all the splices and then coiled the twelve-mile length flat in the hold of a very large laying barge for transport to the site.

In the meantime with assistance from the Puerto Rican power authority, the entire underwater cable route had been mapped out and checked for bottom conditions by divers. Also, triangulation stations were set up on the mainland, on the receiving island, and on one other small island nearer the mainland. By continuously checking the signals from these stations the laying barge would be able to maintain the laid out course accurately. Over this course the mean high-water bottom depth varied between only forty and sixty feet. The shallowness of the ocean at this eastern end of Puerto Rico is in great contrast with ocean depths north of Puerto Rico where soundings have disclosed one of the world's deepest chasms, the Brownsons Deep, nearly twenty-seven thousand feet to the bottom.

The submarine cable arrived at the laying without incident, and the necessary tow tugs and side thruster craft were assembled. It was deemed essential to complete the laying during the daylight hours of one day; consequently the operation was delayed considerably for propitious calm weather conditions. As it turned out, on the advice of the local weather agency given in Spanish and misunderstood, the lay got off to a false start, unfortunately. After the cable had been paid out for less than three hours, a storm started building up rapidly and the operation had to be aborted. Marker buoys were attached to the cable at its farthest reach and the barge anchored there to ride out the storm. It was two days before calm seas returned and the laying could be continued. No further adverse incidents occurred, and the operation was completed successfully.

As is customary with any newly installed cable circuit before it is placed into service, a so called hi-pot test is made under specified conditions. The test equipment is simply an adjustable high voltage direct-current source that is connected in turn between each conductor of the cable and the water. For this thirty-three-thousand-volt cable the passing test required was one-hundred-thousand volts sustained for five minutes. The hi-pot equipment was an older unit supplied by the Puerto Rican power authority.

As the voltage was being run up on the first cable conductor to be tested, when reaching approximately sixty thousand volts the hi-pot equipment itself failed and arced over heavily. This second mishap again delayed providing electric service to the island—this time for almost a week until the hi-pot was repaired and serviceable again. And at that point the real problems with the new cable showed up.

After passing the rebuilt hi-pot tests, the cable was placed in service for the island. Two months later it failed. Fault location equipment located the failure at the second splice from the mainland. This joint section was picked up, laid across a barge, repaired, and again dropped to the bottom. To check the serviceability of the repaired joint the hi-pot was again brought out and the cable retested. This time well below the passing hi-pot voltage there was another failure—again at a splice, the fifth from the mainland. Again the failed joint section was picked up and repaired and the cable retested. This resulted in still another splice failure. With that occurrence the Puerto Rican utility decided that all the splices in the cable must be defective and abandoned any further attempt at making repairs.

At this stage the cable's manufacturer was given notice that a full-length, properly made replacement would be demanded. The AB Company, after reviewing its splice design and after several of its engineers had been in Puerto Rico inspecting the failed line and examining two of the cut-out joints, decided their splices were not defective. They informed the Puerto Rican utility that they believed the failures were in some way related to the operation of their transmission system and that they would assist the utility to determine what this was, but that they would assume no responsibility for the failures.

The upshot was a suit filed by the Puerto Rican power authority, and to assist the case their local attorneys engaged the consulting services of a prestigious engineer firm. This firm, which I will designate as the CE Company, removed a number of the joints from the abandoned cable, took them back to their laboratory, dissected them, and then produced a voluminous report clearly showing the splices were of poor workmanship and seriously defective.

It was just about at that time when I received a call from the cable manufacturer to assist in their defense against this suit. AB Company's attorneys were a Washington, D.C., based firm who made available to me copies of all written documentation, such as recorded interviews, depositions, interrogatories, affidavits, and such. Also, I spent considerable time at the AB plant reviewing their cable and joint design and manufacturing procedures. I talked with the factory splicers and watched them make up several dummy joints similar to those used in the Puerto Rican cable. Finally, I reviewed the factory's certified test reports made on the cable prior to shipment. The full twelve-mile, flat-coiled length had easily passed the new cable test requirement of 125,000 volts (d-c).

From all of this it appeared to me that the cable and its sixteen joints were sound when the shipment left the factory. This would leave the possibility of one or more adverse happenings to the cable either during its transit to site or during the laying operation and the connecting up operation or during the hi-pot testing or during its short time as part of the transmission system supplying power to the island. Checking the records and talking to the people involved, I eliminated the transit to Puerto Rico and the delay in completing the cable laying due to the unforeseen storm. The breakdown and arc over of the hi-pot tester during the first test could conceivably have imposed a high voltage jolt (transient) to the cable, but this would have affected only the conductor under test. The odds were much greater that the cable received some severe damaging high voltage surges from the Puerto Rican systems operation. So that was the area I concentrated on, and as it turned out, I hit "pay dirt."

In the report prepared by the power authority's consultant, although it went into great detail disclosing the cable splices' defective workmanship, there was no statement finding their design faulty. From this I reasoned that although it might be possible for one or even two splices to be sufficiently poorly made to cause them to fail, it was certainly out of the realm of possibility for all sixteen splices to be so poorly made. This then reenforced the theory that some system overvoltage surges, which would simultaneously affect all of the cable and all its splices, had been operative.

Another factor indicating that high overvoltages had been present was disclosed when I visited the laboratory of the CE Consulting Company and was allowed to inspect the dissected remains of the first and second failed joints. Using a low powered microscope I found a whole series of incipients, viz. tiny carbonized spots in the insulations of the cable ends that had been adjacent to both joints. Also, some of these same incipient spots were evident in the insulations of the second joint that failed under hipot testing. (The first joint had failed under service conditions and was too burned up and generally carbonized to detect incipients.) This was a very important finding, apparently overlooked by the CE laboratory technician, since a scattering of incipients usually indicates the insulation has been exposed to extremely high transient voltages.

All transmission systems can be subjected to destructive transient overvoltages under certain circumstances. Lightning striking near an overhead line can cause a traveling wave of millions of volts racing up and down the wires at near the speed of light. The wave can discharge and arc over porcelain insulators normally quite effective for service voltages and if it enters a cable circuit can either cause it to fail or, more likely, to pepper its insulation with tiny incipient carbonized spots. Although every element of the insulation gets the jolt, the wave velocity is so great that at any one spot it lasts only a few hundredths of a microsecond. Another cause of transient overvoltages is inherent in the power system itself. Switching operations on a high voltage transmission line will always generate transients, the severity of which depends on the amount of power being switched and the design of the switch itself.

Although a cable may not fail when jolted by a transient, the scattered incipients can set up the insulation for an early failure under its regular operating service conditions. This is particularly true at joints in a cable, since there is always some reflection of the traveling wave at a joint that augments the overvoltage.

Our attorneys managed to secure detailed installation drawings and equipment specifications of the Puerto Rican transmission system that was set up to supply the island. I studied these for several weeks and made many calculations to determine the parameters of the system as they applied to the propagation of transient overvoltages.

With regard to lightning transients I found ample reasons to expect the submarine cable to have sustained some severe jolts. First, Puerto Rico is in a high thunder storm region. During the two months' service operation of the cable, Puerto Rico recorded three heavy and two light thunderstorms. Second, the lightning arrester protection at the cable entrance terminal was both inadequate in rating and installed with an excessively long grounding wire—which increased the surge voltage traveling into the cable.

With regard to switching transients the situation was even worse. A disconnect switch was installed in the overhead line just before connection to the underwater cable. This switch was used to isolate the overhead section from the submarine cable. Since the long cable retained a great amount of electric charge when disconnected, a second necessary function of the switch was to ground the cable as soon as it was disconnected so that its charge could be leaked off. It turned out that the installed switch was entirely inadequate to handle the cable's large charging current. Whenever the line was switched at this point, instead of cleanly opening or closing, the switch generated a series of arcing restrikes, resulting in recovery of a voltage that built up rapidly to many times the normal service voltage, and this could and undoubtedly did prove very damaging to the cable and its joints.

Thus the evidence was fairly conclusive that this submarine

cable was damaged in service by traveling wave transients and excessive recovery switching voltages and its premature failure resulted from these. I prepared a detailed report replete with traveling wave and switch restrike calculations and submitted it to my client and the attorneys.

That day when I submitted my report in the Pennsylvania Avenue, Washington, D.C., offices of our attorneys is well etched in my mind. I went over the technical reasoning in as nontechnical a manner as possible and showed how the calculations bore out the reasoning. In the middle of the discussions one of their secretaries rushed in. "Hurry!" she cried. "Look out the windows." We all rushed to the windows and, looking down, saw a vast parade of decorated cars and people lined up along the avenue facing the federal executive offices. It was the homecoming of the Iran hostages, and all the government employees and most of Washington had turned out to welcome them home.

Well, we continued our discussions for several hours, with the result that it was agreed to vigorously contest the Puerto Rican suit and base the defense on my report. We also discussed courtroom strategy, recognizing that some, if not all, of the jury might not understand English.

How do you make a lay Spanish-speaking jury understand and visualize a traveling wave transient or a switch restrike? I had an inspiration and went to a toy store where I bought two Slinkies. Most of you will remember what a Slinkie is, I am sure—a four-inch-diameter, flat wire spring of many turns that could "walk" downstairs. With the two Slinkies joined to make a longer one, if it is stretched out and given a shake at one end, a perfect traveling wave will flow down to the other end. If that far end is held rigidly, the wave striking it falls to zero and then is reflected back somewhat diminished. If the far end is held lightly as by a thin suspended string, when the wave strikes it the wave amplitude (voltage) is seen to almost double before reflecting back. Also at the joint (simulating a splice between the two Slinkies) when the wave strikes, it moves to a slighter higher amplitude and sends back a small reflection. These actions by the Slinkie line are almost the same as for a traveling voltage wave in the cable, and I planned to make this demonstration in court at the trial.

To show the switch restrike phenomena, the best I could do was prepare a series of drawings and charts showing stage by stage development of the recurring higher and higher voltages flowing into the cable as the switch began the arcing restrikes.

Eventually a trial date was set and I arrived in Puerto Rico, as indicated previously in this long narrative, two days before we were to go into court. My wife and I were delighted with our elegant quarters in an elegant hotel on the beach. We unpacked quickly and then went down to wander about the hotel and see all of many amenities. We had hardly gotten started when we were spotted by one of the law clerks from our attorney's office. He seemed excited and said he had been looking for me and had some important new information about the trial. Well, it certainly was important; in fact, it was a bombshell!

It seems that our two attorneys and the power authority's attorney had been closeted with the trial judge in his chambers all of the day before and were there again today. A decision regarding whether there was going to be a trial was in the making, and the answer to that was to be given this afternoon. According to my informant, the judge had been studying briefs submitted by the opposing attorneys and was looking for a basis to settle the case without going into court.

It wasn't until late afternoon that our attorneys got back to the hotel with the bombshell. The case was settled! In fact, the power authority's suit was dismissed. The judge, with the evidence of both the power authority's uncertain contentions and our solid technical conclusions before him, had decided that, in fact, the power authority had no case. Moreover, he succeeded in convincing them.

That night we had a celebration banquet at which my wife and I were the honored guests. We toasted each other and regaled the feast with snatches of the legal and technical events that gave our side victory. To overcome some of my disappointment at not having the opportunity to display the demonstration tricks in court, I did them at the banquet—with Slinkies, red and white gloves (to show plus and minus polarity), and switchgear charts. My audience insisted on misinterpreting each demonstration with facetious remarks and laughter, but in the end they agreed they were completely convinced.

The legal victory was great, but my wife and I had come to Puerto Rico expecting to stay at least a week or perhaps longer. We were going to take every opportunity to explore Puerto Rico, to shop leisurely in San Juan's downtown, and enjoy the wonderful facilities of our very unusual beach hotel. Now on our first day of arrival, the "show" was over and there was no real necessity for me to remain here any longer. Very fortunately our inquiry at the airport for a return flight indicated all immediate flights were sold out for the next five days. We made the most of them.

Although I was not involved, I learned that the Puerto Rican power authority ordered cable from a foreign manufacturer and installed proper new lightning arrestors and switchgear capable of handling a high capacitance circuit. And as far as I know, the little island that was powerless is now enjoying the amenities of adequate electric supply.

The Case of the Lineman's Helper

This case in which I became involved illustrates (painfully) the great importance of job training for personnel in any responsible industrial work. And this is especially true for electrical workers in general, but a hundredfold more true for those who work on high voltage transmission and distribution systems. To ply their trade as licensed electricians, the aspirants go through many years of study and training as apprenticed helpers before they pass the examinations and attain the coveted status as "journeymen"—recognized by state labor boards as capable of undertaking electrical work on their own. A journeyman electrician not only knows how to select and install electrical wiring and equipment, but he is also familiar with the safety rules and practices embodied in our National Electric Code and National Electric Safety Code plus any additional state safety regulations.

Since this story concerns those utility electricians known as linemen, let us take a look at their specialized functions. As the name implies, the linemen install and maintain the transmission and distribution overhead and underground lines serving the utility's customers. This work invariably deals with high voltage systems, (four thousand to thirty-three thousand volts in distribution circuits and anywhere up to half a million volts in bulk power transmission). Obviously, to provide the security and safety dealing with field work on such high voltages, linemen receive highly specialized training. Also, retraining and refresher programs are carried on for linemen at regular periods during their employment.

What is said above applies to all large, well-staffed power utilities. But for some much smaller utilities supplying rural communities inadequate staffing is the rule rather than the exception. These entities have no generation of their own but must buy energy from large private or federal facilities. Few have an engineering staff of even a single engineer, and most rely on short-time hiring of a consultant to assist in technical problems. Generally, the small number of electrical workers requires each to have a multitude of duties, with little chance to become properly specialized in any single area. Job training is more or less adequate but does not go beyond bare essentials.

In the Northwest there are many small power utilities serving small towns but also covering many miles of wide open country with their transmission and distribution lines. These are invariably either Co-ops, REAs, or PUDs. As their names imply, these are companies formed by a cooperative effort of a community or, companies supported and to some extent engineered by the Rural Electrification Administration, or companies locally voted into being as a public utility district of a county and subject to county administration.

Several years ago a lineman's helper working for such a county PUD in Washington State sustained a fatal injury under rather odd circumstances. I was called on to investigate this accident, determine what happened, and suggest how to prevent any reoccurrence.

As I learned the bare details, it appeared that there had been a heavy electrical storm the previous evening and night. Early in the morning of the following day, the operator of an irrigated farm called the utility to say that his water-pumping motors had lost power and shut down sometime during the night. To check on this, the PUD sent out a lineman and helper. They traveled to the trouble site in the usual lineman's boom truck, containing spare parts, tools, and instruments to cover most eventualities and a bucket lift.

The PUD's distribution scheme to supply the widely scattered farms was to run their three-phase, 14,800-volt overhead pole line as a feeder along the main county roads and then provide long tap-off lines to the farm properties. These tap lines were also three-phase, 14,800 volts, and could be either on overhead poles or as underground cable. At the farm in trouble, the tap line was in the form of a three-conductor cable. The cable was connected into the main feeder at the pole top and then ran down the pole and entered the ground where it was trenched in all the way to the farm's pump house.

When the PUD's trouble crew arrived at the site of the farm tap-off, they parked their truck off the highway and walked into the farm property. There they examined the large pump motors and control panel and quickly verified that one of the three phases had no voltage. Very properly the automatic disconnects had opened up to prevent damage to the three-phase motors; this would occur if operated only with two phases.

Returning to the highway, the linemen immediately saw that one of the cutouts at the top of the pole had its "gate" hanging down. This told him that the fuse in that particular phase had burned out. The cutout gates in the other two phases were closed; therefore, those fuses must be intact. This then explained why the pump motors had been energized with only two phases and then shut down.

From this finding the lineman reasoned that the lightning storm of the previous evening had sent a surge along the overhead feeder line concentrated on one phase conductor and this had blown that phase's fuse. All he had to do then to restore service to the farm pumps was to replace the blown fuse and close in the cutout. A quick and simple solution and he had plenty of spare cutout fuses in the truck.

At this point I should describe how a tap-off from a distribution feeder line is made. The three conductors of the tap line are not clamped directly to the three conductors of the feeder. Each tap conductor first goes to a fused cutout, the other end of which is clamped to the corresponding feeder phase wire. The fuses are there to protect the system in case a short circuit or heavy overload occurs on the tap line's loads. In addition to the fused cutout there is a lightning arrester or surge divertor across each tap conductor, one end of which is connected to the hot side of the cutout and the other side to the ground wire on the pole. Purpose of the lightning arrester is to bypass to ground any overvoltage surges traversing the feeder line due to lightning, switching, or other causes.

Now the first flaw in the lineman's reasoning can be seen. Obviously if a lightning surge had caused the cutout fuse to blow, the lightning arrester, just ahead of it, was either defective or was also severely damaged by the surge and it would be necessary to replace both the fuse and the lightning arrester to restore service safely. But there was yet another more serious flaw in the lineman's reasoning. What if the surge entering the tap line, having come through the ineffective lightning arrester, having traversed the cutout fuse, then caused a failure in the tap cable itself—at its pot-head terminal or somewhere down in the insulation of the conductor? In either event, such a failure in the cable would represent a short circuit to ground, thus blowing the fuse, and the tap would certainly be in no condition to be reenergized.

Unaware of the above probabilities, the lineman and his helper went about positioning the truck under the pole so that the bucket lift on the truck's boom could be in working location at the tap connections. The lineman then elevated himself in the bucket and removed the blown fuse. After checking the type and rating of this fuse, he called down to his helper to get out a new fuse and tie it on to the hand-line hanging from the elevated bucket. This was done and the new fuse was inserted into the cutout. All that remained then was to close in the cutout to reenergize the phase. Opening and closing a cutout on a high voltage line is done using a six foot so-called hot (insulated) stick with a hook on one end to engage the ring of the cutout switch blade.

The lineman had his hot stick ready, engaged its hook into the cutout, and pushed in vigorously. Instantly there was a blinding arc flash as both the switch and fuse disintegrated and a white hot (five-thousand-to-six-thousand-degree Fahrenheit) arc stream rose rapidly to one of the feeder phase wires that was directly above that cutout location. The moment the arc contacted the feeder line, it melted it in two and the long end, still live, fell to the ground. On its way down it struck the helper on the shoulder and killed him instantly. He had been standing under the overhead line and apparently froze at the roar of the arc blast. The lineman in the bucket, although over six feet from the arc stream, received burns about the face and hands due to heat radiation of the arc.

A week or so after the accident I was able to inspect the site, and a week after that I talked to the lineman at some length. Reconstruction of how it happened was not difficult. The lightning storm preceding the accident day had indeed initiated a surge on the feeder line that entered the tap line. The lightning arrester did not bypass the surge to ground as it was intended to do. Instead the surge continued through the cutout fuse and then caused a flashover at the pothead of one of the tap cable conductors. The flashover at the cable carbonized and burned the pothead insulation and left it short-circuited to ground. When the lineman replaced the fuse in the cutout and closed it in, he virtually short-circuited 8,545 volts to ground of the feeder phase backed by the large power capability of the utility's substation. This is what caused the very high arc current and the rising arc stream.

Of course the lineman should have examined and tested that tap cable before attempting to reenergize the fused phase. He could have seen the pothead damage. Also, he had a "meggar" instrument in his truck for just such determinations of insulation failure. In any event, he did neither of these things.

To verify the lightning arrester's incapability I sent it to an independent laboratory for test. They found that arrester had a design defect and would not respond to properly bypass surges of any magnitude.

One other factor was also responsible for the fatality, and that was the particular mounting of the tap cutouts in relation to the overhead feeder lines. Each of the three cutouts was directly under a feeder line and not more than one foot below it. The fuses generally used and the one in place on that line are known as expulsion type. When a fuse blows due to a heavy current flow the gases generated act like a small explosion. To avoid damage to the cutout body the top of the fuse is made to break open and expel these exploding gases. In this accident, due to the severe short circuit, the expelling gases were in the form of an arc. Therefore, cutouts should never be mounted directly under any line feeder.

The family of the deceased helper, facing a serious financial situation with no wage earner, sought an attorney's advice. They were told that it was not possible to sue the PUD since the helper was their employee and state workmen's compensation rules forbid it. However, suit was entered against the manufacturer of the defective lightning arrester. I became involved in this and submitted a brief that included the independent laboratory test results. This suit did not come to trial. The manufacturer eventually admitted the faulty design of the arrester and was now making and selling a corrected and improved version. A reasonable financial settlement was finally negotiated and the suit terminated.

Before leaving this case I had several discussions with the PUD management. I wanted to air my concerns for this lack of an ongoing lineman-training program and the unsafe mounting of pole top equipment where the feeder lines were being tapped. I believe our talks were constructive and bore fruit. A one-week refresher line safety educational program was instituted for their employees, and this was to be repeated yearly. Also, some of my recommended changes were made in the PUD's line construction practices.

The Case of the Frightened Executives

It was 1943, during the Second World War, and the United States with its allies was at war on three fronts. The nation's industries were turning out vast quantities of war materiel for these fighting fronts—on land, at sea, and in the air.

I was in charge of my company's electrical insulation laboratory at its huge wire and cable plant in New York. We had just received the navy's "E" for excellence in producing magnetic minesweeping gear and the successful degausing ships' cable belts. This latter was the means for neutralizing the magnetic field of our ships, the normal presence of which could set off the German magnetic mines that infested our east coast shipping lanes. I was working during that period with the then Captain Hyman Rickover, who was in charge of the navy program.

Then almost literally a bomb fell on our company. One of our plants in New England had been producing about a third of the army's requirements for field telephone wire, and most of what was made was going to our allies—England and, in particular, the Soviet Union.

At that period of communication technology, radio was of little use on the battlefield due to excessive noise interference and the fact that the enemy had ready access to it. The only reliable means for providing intelligence between the rear and forward fighting segments was field telephone. These instruments, with hand-cranked dynamos providing the necessary electric power, were connected together by so-called field wire—insulated, twisted together, two-conductor cable. These cables are small in size but very strong in tension due to incorporation of a steel wire in addition to the signal-carrying copper conductors. In use the field wire might just be thrown haphazardly on the ground from moving point to point or it might be buried in shallow furrows or ditches.

The wire plant was working around the clock to produce the great quantities of field wire that the army ordered. Their requirements became so large and urgent that the company was under constant exhortation to produce more wire faster and faster.

And so, while the plant's manufacturing and finished wire inspection managers were busy elsewhere (or looking the other way), the wire mill started to ship out field wire without making the twenty-four- and seventy-two-hour inspection tests required by army specifications. After immersion in water tanks for the above time periods, each reel of field wire had to be measured and show an insulation resistance of at least 1 million ohms (one megohm) to be passed for shipment. The ohm is a unit of electrical resistance and in one sense can be looked at as the resistance by the insulation to the escape of current flowing in the field wires' conductors. It was considered that the insulation had to show at least a resistance of 1 million ohms so that (1) use of the field telephones connected by field wire would be an effective communication means and (2) there would be insufficient signal leakage through the insulation to impair operation and, most important, be picked up by the enemy.

Well, after a while, as a result of employee gossip, the army found out what was going on at this plant and immediately the Department of Justice brought violation of contract charges against the company and criminal charges against its officers—its president and vice president and the New England plant managers. The criminal indictment stemmed from the fact that the government charged that defective field wire (that is, wire with less than the specified immersed insulation resistance) endangered the lives of our soldiers and our allies' soldiers on the fighting fronts by causing faulty telephone communication and being subject to enemy pickup due to earth leakage.

Since I headed up the corporate insulation laboratory, the company almost frantically turned to me to help them out of this ugly predicament. You can't imagine what deference I was suddenly shown by the VIPs and executives. They didn't call me up to their plush offices—they came down to my little cubbyhole office to see me and hang on every word as I described what an ohm was, what *insulation resistance* meant, and what this had to do with the army requirements that had been voided. I happened to have a 1 million ohm instrument standard in a small walnut case on my desk. Both our president and VP each held this little box reverently in his hands and kept saying in a hushed voice, "So this is the 1 million ohms they're talking about."

Finally we outlined a plan to lay and bury field wire in several three-mile sections just as might be done on the battlefield, but to have each section with a different insulation resistance—from well over 1 million ohms down to one section in which the insulation had been entirely removed, leaving only the bare conductors. We planned then to monitor each section for communication capability and signal leakage to earth. Of course this required a very large field operation—acres of instrumented ground, trenching equipment, and a good-sized staff, laborers and technicians.

You don't have to guess that I got all I asked for—pronto. A car and a chauffeur were assigned to me personally, and I began scouring the countryside looking for a suitably large, plowable field for our operations. Also I wanted a site near the river so that there would be enough residual moisture in the soil to make the test severe. After a week of hunting, the only area meeting our needs was a large estate (not a farm) with beautifully kept grounds, trees, fruit orchards, flower beds, and lawn and a spacious, handsome nineteenth-century manor house and outbuildings. The owner, a paper manufacturing executive in West Virginia, couldn't afford to live there because of the high taxes. I was sure he would turn us down when I described what we intended to do. But no, he was delighted to take the very substantial sum we offered and assigned us one of his large grassy fields sloping gently toward the river.

Within the next two weeks we installed some eighteen miles of field wire, some laid directly on the surface of the ground but most buried in shallow trenches. Also at strategic places along each run suitable pickup probes were buried connected to a centrally located amplified monitoring equipment station. Then the detailed testing got under way and we were at it day and night for the next month.

Can you guess what we found? Perhaps not. Well, we found that it didn't make a bit of difference in communications efficiency or in signal leakage whether the insulation resistance of a reel length was 10 million or one hundred ohms. Even for the bare, uninsulated section lying in wet ground there was no measureable change in telephoned communicating ability and the signal leakage was so small that even with the best (at that time), most sensitive pickup devices we lost the signal beyond a few feet away from the wire. Also, there is no *mystery* about these results—the nature of voice frequency communication over wires could have predicted it.

With completion of the tests and my report on its findings, our company's attorneys asked for a pretrial hearing, which was granted. The nature and results of our field wire tests were presented in detail before the government's attorneys and experts. As a result the criminal charges against the company's executives were dropped.

Later the company *was* found guilty of a contractual violation, but some mitigating circumstances—particularly the Army's constant pushing for faster production and the laxity of their own army inspectors, et cetera—helped to reduce the fine to a reasonable amount.

The Case of the Untested Switch Box

One of the largest, if not the largest, paper mills in the country is located up in Sitka, Alaska. Started as a comparatively small operation producing newsprint from logs indigenous to the surrounding forests, the mill rapidly expanded in large steps to its current size, producing all types of commercial papers including hardboard.

Like most modern heavy industries, paper mills literally run on electricity—lots of it. There are hundreds of small and large motors and their control gear performing all manner of tasks from turning the giant pulp beaters to slitting the paper rolls. The Sitka plant not only used a great deal of power but also generated most of what it used. All of which indicates that successful plant operation depends greatly on an adequate staff of competent electricians having a good deal of experience in installing and maintaining a wide variety of electrical equipment.

The "Case of the Untested Switch Box" came about when I was called as an expert by a small firm of attorneys in a small Washington town. They had undertaken to represent certain former employees of the Sitka paper mill in a suit against a major

U.S. supplier of electrical equipment. The background was as follows:

Another major expansion of the Sitka plant was under way that required a number of new four-thousand-horsepower pump motors. These had been installed by their manufacturer as the purchase contract called for. To operate and control the large pump motors Sitka ordered a suitable load center switching unit from a well-known, reliable switch gear manufacturer. The load center unit was a catalog item, metal enclosed, complete with roll-out fuses and contactor switches, overload and ground relays, et cetera, and panel instrumentation, all for forty-six-hundred-volt, three-phase operation. This very large unit had to be shipped to Sitka by barge from a west coast manufacturing subsidiary.

Three long-time plant electricians, a foreman and two helpers, all with journeyman ratings, were assigned the task of installing and connecting up the new load center. They had a complete set of prints showing the components, the factory wiring, and the required electrical connections. Also included was a set of instructions for operating the equipment.

Meticulously following the manufacturer's instructions, they installed the unit in its permanent location and made all the external connections. When they had doubly checked everything and were satisfied that all was as it should be, they closed in the forty-six-hundred-volt supply to the unit and pushed in the start buttons. What should have happened was that the contactors in the unit should have banged closed and sent the power out to the big motor control circuit on the mill's operating floor.

What did happen *was nothing*. Repeatedly pushing the starter buttons only emphasized that something was very definitely wrong. So they decided to open the metal enclosure door and make some checks for the presence or absence of voltage at the critical points. Let me here indicate that on all such enclosed equipment the door has an interlock that deenergizes the whole unit when opened. However, a means is provided for overriding this interlock—to be strictly and only used by authorized electricians in case of faulty operation.

Of course they bypassed the interlock and opened the steel door and then, with the schematic wiring prints spread out on the floor, rechecked their power connections to the unit but found no problem there. The prints also showed a number of voltage checkpoints and indicated what readings were to be expected if all was in working order. The electricians had a voltage measuring instrument to make such check readings. This was a small hand-held V.O.M. (volt-ohm-meter) having a digital readout and a flexible pair of insulated connection wires with short metal end prods.

This connection pair on the meter assumed great importance in the subsequent legal action, and I will describe it more completely. The wires of the pair were insulated with a relatively thick polyethylene extrusion, one colored red, the other black, and about five feet in length. At the prod end each insulated wire had an overlying six-inch hard insulating tube joined to the metal prod as a hand-hold. In use, one end of the pair is plugged into jacks on the V.O.M. and the prod ends are touched to the contact points to be checked.

With the load center door open and its interlocks bypassed, as long as the contactor/fuse assembly is not rolled out, the unit remains energized. Facing the front and low down were the three fuses vertically placed in fuse holders. These were very large fuses, each about two and a half inches in diameter and about eighteen inches long. The voltage between any pair of fuses was 4,600 and from any fuse to ground (or the enclosure) 2,655 volts.

Of the five checkpoints tested, the meter readings showed only one in rough agreement with the schematic print. The other four did not agree at all, and two of these showed almost 2,600 volts instead of 240. By the time these check tests were completed, only some five or six minutes had elapsed since the enclosure door was opened.

As the three electricians were crouched in front of the load center studying the instruction manual, the big fuse bank suddenly arced over violently to the partly open, grounded metal door and each man intercepted part of the arc stream. All were badly burned but alive. The plant rescue group came in quickly and by chartered plane flew the men to the burn center in Seattle.

Two of the electricians had been somewhat shadowed by the foreman, and their not-so-deep burns yielded to some extent under cosmetic surgery after six or seven months of hospitalization. The foreman, however, was so severely burned he was in the hospital almost two years before being released.

Nearly three years after the accident, the electricians decided to sue the manufacturer of the load center on the basis that it was a defective unit. However, they found that none of the attorney firms they contacted in the city wanted to take on their case. Generally their position was that only a slim chance for a favorable verdict existed, because the manufacturer was just too well known for its size, technical excellence, and reliability in power equipment and know-how. By chance the small firm of attorneys in rural Washington mentioned previously heard of the case, interviewed the electricians, and committed themselves (and incidentally me) to the case.

Early in our investigation I learned that the manufacturer had sent two field engineers up to Sitka a few days after the accident. They reported that there was nothing wrong with the load center as manufactured and the arc over must have been caused by some thoughtless action of the electricians. And this was the position taken by the company when informed of the suit. In essence they said, "Any idea of defect in the unit was nonsense. The accident was caused by the electricians, who must have draped the V.O.M. connection pair from their test meter over the forty-six-hundred-volt fuses—thereby shorting them and drawing the arc."

As the case developed I was able to carefully examine the load center switching unit or what was left of it after the heavy arcing. Many of the numerous relays, although they were in a side cabinet, were scorched and carbonized—as were also much of the wiring. On comparing the schematic wiring points in the instruction manual with the as-wired state in the unit, I found that the starting relays had been installed backward and the central circuit was misswired. Also, and most important, no part of the wiring circuits had a grounded point (a connection to the steel enclosure) as called for by the schematics.

More clarification as to what happened came as I read the many memoranda and depositions of the various insurance and engineering investigators. Also, I had the opportunity to interview the three injured electricians at length. They assured me that at no time during their taking the voltage check readings had the meter cords inadvertently touched anything in the unit. Also at the time of the arc over the V.O.M. was not being used; they were crouched together studying the schematic prints spread on the floor.

My examination of the V.O.M. and its flexible connection pair definitely discredited the manufacturer's contention that it had caused the accident. I found the meter with its connection pair in perfect working order. There was a small amount of carbonization on the insulation and rigid tube of one wire of the pair, but nothing to interfere with the meter's operation. Had any part of the meter's pair caused a short circuit of the forty-six-hundred-volt fuses, that part would have been vaporized instantly and the meter itself burned badly.

As a final note, we had good reason to believe that this load center had never been tested by its manufacturer.

Our request to the purchase department of the Sitka mills disclosed that no certified test report concerning this load center had been received. When we requested the manufacturer to produce a copy of its test report, we received the answer, more than a month later, that they couldn't find it. Their excuse was that part of their west coast plant had been moved east and some of the records had not turned up yet (nearly three years later). As it turned out, we never did receive a test report and what with the miswiring we found, the inference was pretty clear that this unit had been shipped untested.

Well, the case came to trial in U.S. District Court in Seattle. The defendant, the manufacturer, had the manager and half a dozen of his switch gear engineers in court as expert witnesses. I was the expert witness for the plaintiffs, the three injured men. As a surprise the defendant had a whole, new load center switch unit, supposedly a duplicate of the Sitka unit, set up as an exhibit in the courtroom.

As it developed, the arguments centered about the reality of two divergent claims. The plaintiffs claimed that the load center as delivered to the Sitka plant was defective and it was this defect that caused the accident. The defendant claimed that the load center was not defective and that the accident occurred as a result of a short circuit caused by the plaintiffs.

In presenting their defense, the defendant's experts gave a detailed account of the load center's design background, the quality control exercised during manufacture, and the hundreds of units currently giving satisfactory service all over the world. As for the way the accident occurred, they showed how easy it was to drape test wires over the exposed fuses in the unit set up there in court—implying that if the fuses were energized at forty-six hundred volts, arcing would certainly follow.

It fell to our side to offer the rebuttal to their presentation. First we set out the facts in our attempt to obtain a copy of the presumed factory tests on the load center and our forced conclusion that no tests had ever been made. Then with the original V.O.M. and tests leads as exhibits in my hands, on the witness stand I showed conclusively why these could not have taken part in any short-circuiting of the fuse bank. Even if the test leads had been draped over the forty-six-hundred-volt fuses I explained why they could not possibly have caused a short circuit or arcing. The polyethylene insulation on the test leads, fifteen-thousandths of an inch thick, could sustain more than three times a forty-six-hundred-volt contact without any trouble. I then offered to drape these same test leads over the *energized* fuses with my own hands to prove my statements.

I was then asked to give my opinion how the accident did happen. I described our findings of misconnections and reversed relays and the electricians' readings of excessive high voltage in the normally low voltage control section of the load center. This high voltage came on when the start button was first pressed, and since there was no grounding connection, the control circuit fuses couldn't blow open. The result was that for five or six minutes a great deal of sparking discharges was going on in the control cubicle, which produced much ionized gassing. The ionized gas, highly conductive, reached the bare forty-six-hundred-volt fuses and there caused them to arc over to the metal door.

While I was giving the above evidence, I noted that the manager of the defendant's switch gear group was paying very close attention. Also he conferred with their attorney several times during my following cross-examination.

Early the next morning when the trial was resumed, the defendant's attorney offered to settle the case for $150,000. Our attorney refused. Again later in the same day another offer to settle was made, this time for $250,000. We conferred on this and decided to refuse. Finally the case went to the jury. They deliberated not more than a couple of hours and then handed down their verdict that the defendant was guilty of negligence in supplying a dangerously defective piece of electrical equipment. They awarded the electricians damages totalling eight hundred thousand dollars.

The outcome of this case certainly made me feel good.

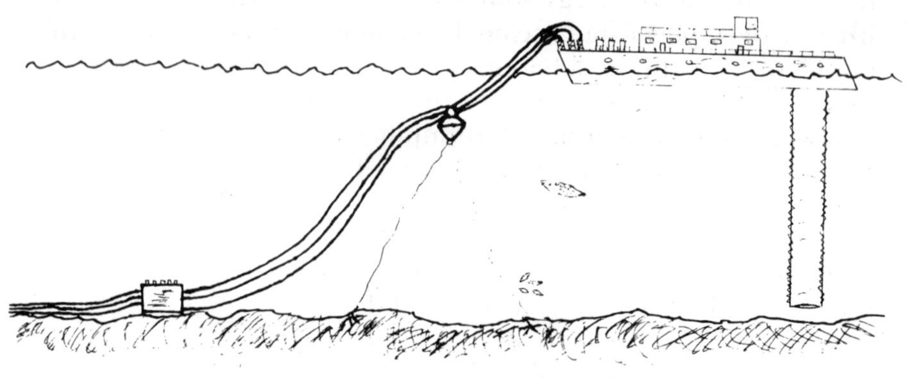

The Case of the Ocean Energy Gamble

In the early 1970's this nation as well as other industrial nations of the world faced a serious energy shortage as a result of an embargo of Middle East–produced oil. Here in the United States it was a scary business while it lasted, with long lines of cars and trucks at every service pump still having some fuel.

An important consequence of this fuel shortage experience was the public's widespread realization that the largest proportion of our oil requirements depended on foreign sources that could be shut off at any time. "Make the United States self-sufficient" was the cry, and both public and private agencies responded with new emphasis on alternate possible energy sources not dependent on oil or coal.

Some spadework had been undertaken by the federal Department of Energy (DOE) for several years. Faced with a shortage in fossil fuels, very limited hydroelectric sites, and the rising costs and serious environmental objections to nuclear power, the DOE had been sponsoring research, for example, in wind power, tidal energy, geothermal energy, and solar energy. Other agencies of the Department of Commerce had been working on conversion of our immense resources of coal shales into liquid fuels.

It was no coincidence that most of the studies sponsored by the DOE involved energy sources considered pollution-free. Utilization of superheated steam from molten rock in the earth's bowels or from direct radiation of the sun or the mechanical forces of moving sea waters requires no burning of fuels and so no atmospheric pollution or ash dumping. However, these projects had been moving along slowly, almost leisurely, with no real urgency to financially support any kind of a major effort. But the situation changed completely immediately after lifting of the oil embargo. Now the White House and the Congress looked to commit substantial funds to research and development in these alternative energy resource areas.

The major push was in the area of solar energy for space and water heating, and solar heat collectors blossomed over buildings, dormitories, residences, and swimming pools and on many structures where weather and latitude were propitious. However, the use of solar energy to provide electric power in bulk amount was far more ambitious and no practical means to accomplish this could be foreseen. The main problem lies in collecting it. It is so disperse; on the average there is less than one watt of useable power per square foot of collection area and its amount is drastically affected by location and by seasonal, daily, and even hourly weather changes. Under ideal conditions a city the size of Spokane was estimated to require a solar collection field of almost one hundred square miles to meet its demand for power—with no assurance of doing this on any continuing basis.

However, an old concept of using sun-warmed sea water to drive an electric generator was revived. This avoided the big collection problem of ground-based solar systems and had other advantages, which aroused great interest. The oceans have already collected vast amounts of solar energy and continue to collect it on the grandest scale. Also, weather, seasonal, and daily changes have little or no effect.

Briefly, the solar sea power scheme depends upon the difference in temperature between the surface and the bottom waters. The surface warm water pumped into a thermal engine using a low boiling fluid such as ammonia vaporizes it, boils it, and increases its pressure and this turns a turbine driving an electric generator. The spent exhaust from the turbine is pumped to a condenser, which is cooled by the cold waters brought up from the bottom, and this reliquifies the ammonia ready for the next heating cycle.

There are many areas in our tropical seas where the surface waters maintain themselves at close to eighty degrees Fahrenheit. At the same time the inflow of heavy cold water from the arctic regions maintains bottom waters close to forty degrees Fahrenheit. This stable forty degree Fahrenheit temperature difference is sufficient to operate a low-temperature–designed turbine. Of course typical land-based, fuel-fired turbines operate at temperature differences between source and sink of many hundreds of degrees centigrade and achieve thermal/mechanical conversion efficiencies of 35 to 40 percent. For the solar sea power engine with its low temperature operation a maximum efficiency of only 3 percent could be expected. Therefore, for a given kilowatt output, the unit's "boiler" and condenser must process more than ten times as much heat as for a conventional boiler and this means processing literally enormous volumes of water. But there is that water—in unlimited quantity for the taking—at "no" cost.

In early 1979 Congress passed a bill, S-1830, to provide research, development, and a demonstration program to achieve the early commercialization of Ocean Thermal Energy Conversion (OTEC) systems. The goals set were to achieve a working one-hundred-megawatt electrical plant by 1986 and a five-hundred-megawatt plant by 1989. The bill provided a sum of $25 million per year to carry out the act, plus $15 million for the initial work on siting and designing the one-hundred-megawatt unit. This was to include not only the thermal engine/generator design but also the plant's platform or ship and the means of bringing the generated energy to designated shore stations.

The siting study uncovered at least two locations close to U.S. shorelines that promised a steady twelve month temperature differential of 40 degrees Fahrenheit. The closest, only 1.7 miles off the southwest coast of the Hawaiian island Oahu, showed a bottom depth of six thousand feet. The second site, about twelve miles off the coast of Puerto Rico at Punta Tuna, was closer to five thousand feet in depth. Both of these sites were considered feasible, and both were included in the ongoing study program.

To carry out the OTEC program the DOE set up several advisory and surveillance committees to farm out the research and development projects and closely monitor progress in all areas. It was a very large undertaking with almost all of its elements outside of state-of-art experience. Specialists from the government and universities and high technology companies were brought into a series of discussions. These led to authorized assignments and funding contracts covering the separate elements that would lead to the realization of a one-hundred-megawatt OTEC facility.

Assuming a successful design and construction of the low-temperature turbo-generator and accessories, there still remained a number of formidable problems—amongst a host of lesser needs. Design of the plant-ship and means to keep it on station under all weather conditions once it is on location is obviously of paramount importance, since the electrical output of the plant must be delivered to shore via bottom-lying submarine cables. The navy undertook this problem with an original plan to use side thrusters activated as needed by radio signals from a land-based Loran coordinating station. However, their most optimistic estimate called for the plant-ship to move around in a so-called watch circle of at least five hundred feet. Another major problem has to do with bringing up cold water from the bottom. Imagine a tube of some kind twenty to twenty-five feet in diameter extending vertically six thousand feet from the ocean bottom to its surface through which the cold water must be brought up without gaining any significant temperature rise. Imagine the stresses on such a giant pipe as it rises to within five hundred feet of the ocean surface, where all the turbulence and wind-driven wave action take place while at the same time it is joined to the plant-ship, with the latter's own considerable wander.

Another major problem, the one with which I became associated, had to do with getting the OTEC power to shore. I spoke about a submarine power cable lying on the sea bottom providing a connecting pathway from a shore substation out to the plant-ship. Then at the sea end of this cable it would have to be joined to some type of flexible *riser cable* going up six thousand feet and there be connected to the output terminals of the plant-ship.

Research and development of the OTEC cables was contracted to a cable manufacturer with long experience in submarine cable technology. The company very soon set up an advisory committee of experts in cable science, materials science, hydrology, corrosion, mechanical and electrical testing, and so forth. I was asked to join this committee, and I served on it until the project was terminated a few years later.

Let us look at what is required of this submarine cable link and divide it into two segments, because each has totally different environment problems. The two segments have this much in common—they will carry one hundred megawatts at 138,000 volts. The fixed segment would rest peacefully on the ocean floor and extend from the shore out as far as about twelve miles (the Punts Tuna site). Existing cable technology appeared able to provide a design for this cable—with one important exception, namely, its working depth of up to six thousand feet. No power cable had ever operated at that depth before. At six thousand feet the water pressure approaches twenty-four hundred pounds per square inch. This enormous crushing force, for example, on oil-filled 138-KV cable would force it to also pressurize its oil channels to twenty-four hundred pounds per square inch. No oil-filled or any other high voltage cable has ever been operated at more than two hundred pounds per square inch.

But solving the pressure problems of the bottom-lying fixed cable segment was simple as compared to the far more serious problems of the riser cable. In view of this, we ignored the fixed cable needs and concentrated entirely on the riser cable. Again let us review the special requirements. It must rise from a junction box at the end of the fixed cable almost one and a quarter miles up to terminate on the floating plant-ship. Taking into account its rise curvature and a possible looped buoying configuration for suspension, its length can approach ten thousand feet. The "moored" plant-ship cannot occupy a fixed position but must be allowed a watch circle for movement—the navy said five hundred feet minimum. Moreover, the top six hundred feet to the surface is affected by currents and storms and these will add proper motions and violent three-dimensional movements to the plant-ship, with correspondingly violent stresses to the riser cable passing up through that layer to its terminals. Thus the design of the riser cable itself and its mode of suspension are the most critical problems related to delivery of OTEC's energy. In such cable the high voltage (138,000 volts) electrical requirements and the mechanical problems are directly dependent. Mechanical integrity must be assured in order to maintain electrical integrity.

Our cable advisory committee met frequently to consider the ramifications of electrical and mechanical design, prototype testing, and cable suspension methods. As individuals or groups we were each assigned to work on specific elements of the project. The apt, for that time, motto of the committee was "OPEC no, OTEC yes."

Two types of cables were to be considered. The first was an oil-filled paper, lead-sheathed type, for many years the transmission cable workhorse, but with special armoring. The second was a so-called solid type cable in which the insulation consists entirely of polyethylene resin and contains no fluid of any kind. Ths type of cable had very little background of experience in high voltage submarine service. My assignment was to design the oil-filled cable prototype. Known as a SCOF (self-contained oil-filled) cable, its insulation is composed of special paper tapes impregnated with a light insulating oil surrounding the hollow central conductor filled with the same kind of oil and maintained at a given positive pressure by external means.

Now the plant-ship was to be designed to generate its electric power at 13,800 volts *3-phase* and then to pass this through step-up transformers to provide terminals for transmission power, at 138,000 volts, still *three-phase*. This means that three riser cables, one for each phase, would be required. Or if the riser could be designed as a three-conductor cable then only one cable would be needed. With regard to insulating quality and electrical performance a three-conductor 138-KV cable can be designed to be just as reliable as one-conductor cables. Obviously a single three-conductor cable would be substantially less expensive both as to cost of cable and of placing into service. However, there were overriding considerations that ruled it out. Its weight (in sea water) was fourteen to eighteen pounds per foot versus seven to eight pounds per foot for one-conductor cables. This was a very important factor considering the necessary suspension of perhaps ten thousand feet of cable. A second factor was the possibility of a failure in the three-conductor riser cable, which would put the whole facility out of commission. With three one-conductor cables and a fourth as spare, failure of one cable would not interrupt the flow of three-phase power to the shore station. And so we settled on single conductor riser cables and I and several others set to work on its design.

It took me about a month to finalize my concept of a modified oil-filled cable to meet the severe requirements set up by our advisory committee. I should point out here that the tremendous crushing force on the bottom-lying cable due to the six-thousand-foot head of water does not exist for the riser cable. That is so because the weight of the oil column in the vertically suspended riser cable increases in exact proportion with depth as does the sea water. If the density of the oil was exactly the same as sea water, there would be absolutely no pressure difference between the inside and outside of the cable. But sea water is somewhat heavier than cable oil, so there does occur a small differential pressure on the outside of the cable. However, even at six thousand feet this pressure can easily be taken care of in the cable's mechanical design.

Toward the end of the 1980 summer, several of the cable designs were set up for manufacturing short lengths to be used for repetitive mechanical stressing tests. By the end of that year substantial lengths of the solid type and the oil-filled (SCOF) cables were being manufactured for extensive electro-mechanical testing in plant-ship simulated configurations. Results of design, manufacturing, and tests on the year's work on riser cable development were presented by the company at a national energy technology conference in Hawaii in December 1980. This was received very well. However, in general the many presentations by the navy and contractors concerned with OTEC development indicated, at least to me, that the national program with regard to goals and accomplishments had, in the vernacular "bitten off more than it could chew." For one thing, the goal of a one-hundred-megawatt power unit had to be scaled back to forty megawatts, still with no assurance that that could be attained. Stationing the plant-ship under all conditions, notably in the one-hundred-year storm of record, also seemed extremely difficult—certainly not solved at that time. The mandated cable life of thirty years looked like it would have to be cut back to ten years. And so it went with almost all elements necessary to realize a commercially sound OTEC facility. Also, by this time the price tag for the OTEC project was approaching $30 million and the DOE was certainly looking thoughtfully at further funding requests.

I spent a few delightful months with my wife in Hawaii looking over the previously selected OTEC site off of Oahu's Waianae-Makaha shoreline. Also, I was working with a Honolulu based consulting firm that was interested in the installation of the OTEC cables—a formidable undertaking.

Through the next year I had very little to do with the OTEC project. Articles on OTEC and its great promise, which had appeared earlier in the technical press, had dwindled away to almost nothing. The great material and logistic difficulties had slowed progress, and this appeared to cause some reluctance to adequately fund further efforts. By 1984 some demonstrations of small prototype low-temperature heat engines had been made successfully, but no OTEC facility of any size had been deployed.

And so the OTEC project and all of its design and experimental fragments were put to rest in the files of those of us who had worked on it. One day, no doubt, when a greater need for self-sufficiency confronts the nation, OTEC may come alive again.

The Case of the Hawaiian Hiatus

For twenty years following my graduation as an electrical engineer I was steadily engaged in research and development studies of electrical insulating materials and insulated cables. As for the cable's conductor, we took it for granted that copper was universally the metal to use. It had gained this position due to the fact that refined copper conducts electricity far better than any other metal, with the single exception of silver. Silver, a precious metal, has a somewhat higher (108 percent) conductivity than copper but cannot be used for wire and cable due to its much greater cost. But other metals are considerably lower in electrical conduction ability. For example, the best of these, gold, has only 74 percent of copper's conductivity and little mechanical strength and is far too expensive for the remotest consideration. Aluminum, the next highest, has 61 percent the conductivity of copper, while iron is down to 17 percent, steel even lower, and lead is less than 8 percent.

On the other hand, aluminum, although having only 61 percent of copper's conductivity, does weigh less than one-third, has almost half its strength (when hard drawn), and is so plentiful that under almost any market condition it costs less than one-third of an equal weight of copper. Adding up these factors, it turns out that bare aluminum conductor can be manufactured and installed in overhead electrical line construction and provide equivalent performance to copper at considerable less cost. And that was the way the power companies with their long transmission and distribution systems were looking at it in the early 1950s. It was a time then for making sure that aluminum could really be relied on to replace copper under the varied conditions of exposure and topography. Power companies were willing to cooperate with suppliers, installing aluminum experimental lines under differing conditions, these to be monitored for performance by both the utility and the wire manufacturer.

By and large the aluminum lines they installed performed very well—as well as copper. And in the heavy industrial areas, for example, around Pittsburgh, Chicago, and the Jersey industrial complexes, aluminum conductors outperformed their equivalent copper, as the latter is corroded to some extent by sulphur-containing gases—always part of the brew in such areas. However, along the seacoasts where the transmissions were exposed to a wet, salt-laden atmosphere, aluminum wires were found to suffer corrosion, in some cases severe. The reason, not long in forthcoming, was that the saline moisture destroyed the normally present hard, protective oxide film on the aluminum surface, exposing the vulnerable metal itself to chemical and electrolysis attack.

That was about the situation when I joined the Kaiser Aluminum & Chemical Corporation in 1950 and was shortly transferred to the Trentwood, Washington, Reduction Plant and Corporate Laboratories as supervisor of electrical research. A major company objective at that time was to prove in selected aluminum alloys as less costly, more durable alternatives to copper conductors, both bare and insulated, and to galvanized steel sheeting for use as roofing, siding, and its multitudinous applications in agriculture.

My section, of course, worked on the electrical conductor problem, and in this we were assisted by the metallurgical research section. Not only had aluminum wires shown corrosion susceptibility in seacoast locations, but also difficulties had shown up in making reliable electric connections between aluminum and copper and indeed between aluminum and aluminum. How we, and of course others, solved the connection problem is an interesting story but will not be a part of this telling. The narrative thread I will pursue here is our solution to the seacoast corrosion problem that led us to the islands of Hawaii and my lifelong love affair with them.

It is one thing to make tests on antiseptically prepared samples of various aluminum alloys in known controlled cabinet atmospheres and quite another thing to set up a set of full-sized transmission lines in a severe natural environment. The preliminary laboratory work is essential, but the final decision must await the results of well-designed and monitored field trials.

Some of these earlier corrosion-pitted aluminum lines installed by power companies were on the west coast where the pole lines were close to the ocean shore. Therefore I contacted the Pacific Gas & Electric Company, San Francisco, and together with their engineers we set up a test site at Bodega Bay. Here five three-hundred-foot spans of various alloys of aluminum conductors were installed on P.G. & E's pole lines not more than fifty feet from the water.

After fourteen months of weathering and monthly on-site examinations we learned some important things, but in general the tests were not as conclusive as we had hoped. One of the experimental lines was made of wires from specially (and very expensively) prepared aluminum rod that was within one ten-thousandths of 1 percent pure. This line showed not the slightest blemish throughout the test period. All of the other conductor lines showed blemishes of one sort or another, but none had any significant corrosion attack. Obviously the selected site was not severe enough. It could take years to develop a corrosion pattern. What we needed was such a severe natural environment that in less than a year if one of our experimental alloys remained intact we could safely predict it would provide a normal lifetime of satisfactory service.

At a technical meeting a few months earlier I had listened to a paper given by an engineer from another aluminum company who was describing their attempts to introduce aluminum conductors on the Hawaiian Islands. After several years of work on Oahu, they concluded that aluminum was incompatible with its very severe salt-laden atmosphere, particularly on the windward side.

So I was convinced Hawaii was the right place to set up our aluminum conductors field study. But Hawaii was an ocean away, and the much greater expense and time that would be required for these tests would be very difficult to sell to management. However, at this point an unexpected occurrence changed the situation completely and I was on my way to the islands within a week after closing out the Bodega Bay tests.

Mr. Henry J. Kaiser, the dynamic chairman and chief executive officer of the vast Kaiser enterprises, had been ailing. His physicians and wife, a trained nurse herself, had been advising a more relaxed life and a move to a climate that was moderate year-round. Finally he gave in to their arguments and the Kaiser household was transplanted to Oahu, to a sumptuous oceanfront estate north of Diamond Head. But any idea H.J. may have had about relaxing was swept away when he looked around his new homeland. He saw a land of great beauty, with gorgeous flowers and trees, an unsurpassed climate, and every natural amenity of breathtaking seascapes and mountain views and peopled by a handsome polyglot native population who were happy, peaceful, and friendly. Yet for all this combination of virtues so sought after by tourists and travelers, the facilities, as to transport and accommodations and publicity, were completely inadequate—in fact, they were in a frozen state. He set about doing something to unfreeze this situation.

He saw red rusted "tin" roofs and sidings on houses and shops and rusted flumes, troughs, and other agricultural paraphernalia everywhere in the sugarcane and pineapple fields. He visited the Honolulu grocery stores and could not find any aluminum foil on the shelves. He looked at the Hawaiian Electric Company's overhead distribution system on poles all around the island and saw only copper conductors. He wasted no time in doing something about these.

At that time in the fifties the only commercial communication link between Hawaii and the mainland was radio telephone. Mr. Kaiser ruffled the "ether" waves considerably calling to all of his aluminum company executives. He wanted to know why Hawaii had been bypassed and overlooked and was not now partaking in the "new world" of aluminum products. He told them to round up some top aluminum sales personnel and engineers and send them over to Hawaii to report to him. With regard to household aluminum foil, he asked that product manager to load up a DC-4 with as many boxes of foil it would hold and send that to him immediately. (I heard later than Henry J. himself distributed these boxes of foil to all the large food stores in Honolulu.)

That was how I first came to Hawaii—to be given an assignment by the big boss to do something that I had been scheming to do for weeks. Mr. Kaiser asked me to meet with the Hawaiian Electric Company engineers, discuss the technical and economic merits of aluminum electrical conductors, and, if possible, arrange for some cooperative aluminum transmission or distribution line tests.

I did this and almost instantly met success. HECO was not only willing but almost anxious to prove in aluminum conductors. Added to the much higher price of copper conductors as bought on the mainland there was the substantial shipping costs with this very heavy metal. They were willing to set up several of our test sites on Oahu's windward shore and would assign two of their engineers to work with us on the project. They did caution me concerning previous attempts by others to use bare aluminum wires on pole lines with no success. I told them we knew about those failures but, based on our own tests, were optimistic that we could come up with a conductor equivalent to copper and having a recipe that resisted saline corrosion.

Back at the Trentwood laboratories we spent the next several months designing the Hawaiian tests and selecting the most promising alloying compositions for drawing into wire. Again there would be two control lines, one made from the currently standardized electrical conductor grade aluminum (EC) and the second carefully fabricated from pure aluminum again (containing less than one ten-thousandth of a percent impurities).

At this point the reader needs to know something of the properties of aluminum as affected by its inclusion of other metals, either by design or simply as impurities. Since it is to be used as a conductor, the aluminum wires should have the highest electrical conductivity possible. For example very high purity aluminum—99.999 percent pure—has a conductivity of 65 percent that of copper. However, pure aluminum has very low tensile strength (about nine thousand pounds per square inch) and stretches inordinately—certainly that is not for overhead tensioned line service. However, only a few tenths of 1 percent of certain other metals raises the wire's tensile strength considerably, which when hard drawn can reach almost to thirty thousand pounds per square inch. This was the type of aluminum conductor, with not less than 99.5 percent of aluminum, that had been standardized for overhead pole line systems. But the presence of that half of 1 percent of other metals dropped the electrical conductivity to 61 percent, and this was the accepted tradeoff to achieve the higher strength needed. The perfect test result on the high purity line at Bodega Bay led us to believe that it was the very small nonaluminum metal content (which at that time was standardized at four-tenths of 1 percent iron maximum and one-tenth of 1 percent of silicon) that caused the corrosions experienced in saline atmospheres. And we had a strong suspicion that iron was the corrosion factor.

Based on the above, the test lines for Hawaii, other than the two controls, consisted of a series of aluminum conductors all with 99.5 percent aluminum content but with varying ratios of iron and silicon making up the additional half percent of composition. In setting up these iron to silicon ratios two properties of the finished conductor had to be maintained, namely the conductivity had to be at least 61 percent and the tensile strength at least twenty-seven thousand pounds per square inch.

It took us about five months to prepare the various alloys, test them, and have them drawn and stranded into conductors. Then we shipped all the reels, there were seven of them, to the Hawaiian Electric Company to await our coming.

It was a different Hawaii for me this time. First, I had my wife, Louise, with me. Second, Mr. Kaiser had acquired a native type hotel property in Waikiki, renamed it the Hawaiian Village, and was busy putting it into first-class shape. My wife and I and the staff we brought along from Electrical and Metallurgical Labs were all put up in the new Hawaiian Village. It was a most interesting place consisting of a number of thatched-roof bungalows surrounding a large round swimming pool—over which a polished floor could be slid for nighttime dances. The lobby building had cane-and-basalt-block walls and a thatched roof and housed a restaurant and entertainment stage. The style could best be described as luxurious, primitive, and native Hawaiian.

For the next three weeks we stayed at the village, sleeping and taking our meals there. Nightly at dinner and after, a two-hour stage show was featured in the restaurant with Polynesian singers, steel guitarists, dancers, acrobats, sword swallowers, and so forth. But the feature event was always the singing of Alfred Apaka, with his magnificent baritone.

Together with the two Hawaiian Electric Company engineers and their crew of linemen we set up the aluminum conductor test lines at two windward locations on Oahu. One was along the shore of Kaneohe Bay, the other farther north near the town of Kaaawa. Both sites were exposed to the steady moisture-laden trade winds blowing across the ocean's vastness onto the island's shoreline. The test sites were activated early in 1952 and were terminated nearly three years later.

As the study progressed and the severity of that exposure became more and more apparent, additional features were added. For long-span high voltage transmission lines where the conductors must withstand very high mechanical tensions, a type of conductor designated as ACSR had been developed. As the initials imply, this was made of aluminum strands surrounding a core of extremely high strength reenforcing steel wires—the aluminum carrying the current, the steel carrying most of the tension. It was well known that galvanic action between aluminum and steel in the presence of saline moisture caused the aluminum to sacrifice itself and go into solution—eventually destroying the conductor. To offset this, the steel wires were generally heavily galvanized. This was effective but not completely so, and efforts to improve the performance of ACSR had continued. We added to our tests ACSR conductors with (1) the usual galvanized steel core, with (2) a new process of aluminized steel core, and with (3) both of these types greased with different kinds of heavy lubricants. Also, several cable connector manufacturers asked to partake in these tests with some of their commercial and experimental products and these were included.

During the course of the tests, we flew out to Hawaii about twice a year to examine and photograph the lines. One can't imagine a more pleasant occupation! On each occasion we stayed one to two weeks. My wife was with me on most of these trips, and several times our children came along. We usually rented a housekeeping apartment somewhere in Waikiki close to the beach we patronized for a daily swim. Of course we made a number of long lasting friendships with Hawaiian Electric Company people and the previously established Kaiser sales office personnel. Through these and on our own we met many other interesting people at the University of Hawaii and the Bishop Museum.

The small companies supplying electric power on the other islands of Hawaii: Kauai, Maui, Molokai, and Hawaii, of course knew of the tests underway on Oahu. We were invited to visit each of these islands and to advise the local utility on what we hoped to learn and what recommendations we had for them. Again lasting friendships were made with people on these so-called outer islands.

Carrying on the conductor tests with the essential help of Hawaiian linemen was a pleasure. They were a happy people, chatting and laughing as they went about the tasks of putting up and taking down lines. It was hot work in the tropical sun, but we had plenty of cool, delicious, coconut milk to allay thirst. One of the men would scramble up a tall coconut palm and throw down half a dozen big coconuts. (We all had on hard hats.) Then with one expert slash of a machete the husk was neatly severed and the halves passed around as drinking cups. Also with the plentiful cultivated fields close by we all had plenty of delicious ripe pineapple—also neatly sliced by machete.

We were about one year into the conductor test period when the answer to seacoast corrosion of regular electrical conductor grade aluminum was unmistakably found to be its iron content. Conductors having decreasing amounts of iron from the standardized four-tenths of a percent showed greater and greater resistance to corrosion pitting. We finally settled on a composition of 99.5 percent aluminum, 0.35% silicon, and 0.15% iron as the best compromise between conductivity, tensile strength, and good corrosion resistance. With the high iron content, the pitting started at the surface of the wire due to galvanic action between aluminum and iron atoms and then went in deeper and deeper.

The tests also showed that ACSR conductors with aluminized

steel core wires were more resistant to corrosion than the standard zinc-coated core. Part of the reason for this was apparent in the relative softness of zinc galvanizing with resulting rupture of the protective coating due to strand rubbing. The aluminized coating was much tougher. Also, the connectors and other aluminum line accessories under test disclosed some weaknesses of materials and design that needed prompt attention by their makers.

All in all, these Hawaiian tests were a boon to aluminum use in the electrical industry. Our manufacturing and sales divisions wasted no time in taking full advantage of our work. Hawaiian Electric Company and the other island small utilities gradually converted all their transmission and distribution lines to aluminum.

Although we have revisited Hawaii many times, my wife and I look back with nostalgia on the first years we *malahinis* (newcomers) spent there in the early 1950s. How different it was then from what it is now as I write this, almost forty years later. Flying safely into the Honolulu Airport in the old underpowered DC-4, you sighed with relief and then with delight as you walked into an almost completely open-sided terminal with thatched roofs and bamboo/cane walls and uprights, the clerks all wearing either color-splashed aloha shirts or grass skirts and beautiful, so fragrant leis of ginger, pekaki, and plumaria blossoms—and overall the hauntingly sweet smell of pineapple everywhere. Today's Honolulu Airport is a vast jungle of concrete and steel, full of hurrying people and overall the sour smell of high octane fuel.

We remember the streetcar rides in Honolulu and out on Kalakaua Avenue through Waikiki to Kapiolani Park and its fine zoo, the gorgeous flowers and blossom-filled bushes, the flame trees, and the magnificent monkey-pods and banyans. We remember Waikiki as it was then, wide open, lots of empty corner lots, very few hotels, no high-rises, and the wide beaches and Diamond Head dominating everything. Now there are a hundred immensely tall apartment houses, condominiums, and blocky hotel buildings jammed together and turning the streets into narrow canyons filled with buses and passenger vehicles from which the beaches and Diamond Head can no longer be seen. We remember walking along the Waikiki beach from the Hawaiian Village out to Kapiolani Park, sometimes on sand but mostly in the shallows, eating mango and papaya salads under the great banyan tree at the old Moana, dancing in the sumptuous Royal Hawaiian Ballroom, visiting the beachfront restaurant that Charlie Chan frequented, and enjoying their wonderful mahimahi topped with macadamia nuts. We remember standing at the top of Nuhuanu Pali, looking down the twisting rough road to the sleepy town of Kailua perched above the sparkling ocean, driving along the old Kamehameha highway to the Blow Hole and the magical beauty of Hanauma Bay, and climbing to the top of Diamond Head and peering out of the old gun emplacements at the stunning views of islands and ocean. We remember standing amongst the crowd on the Aloha Pier at Honolulu's port with the beautiful white Matson liner *Luraline* jammed to the rails with tourists waving good-byes and as the great ship inched away from the dock the band near us struck up, and we all sang, the sad but lovely Hawaiian Farewell Song while hundreds of colored streamers connecting ship to shore started breaking apart as the parting passengers reverently tossed their leis into the ocean as a promise to return.

We remember flying out to the other islands on the Aloha Airline's little DC-3s in which each cabin window had a small slide-up opening for your camera lens and the two Hawaiian stewardesses wore leis and grass skirts and did a hula in flight, the little airport on Molokai, and its one quaint, primitive town of Kaunakakai, site of Hilo Hattie's famous rendering of "The Cockeyed Mayor of Kaunakakai." Here was the Kalaupapa Pali overlooking the still operated leper settlement founded by Father Damien, and here too were our good friends the Yamashitas, who ran the tiny local electric system and invited us to spend some delightful days at their "vacation" cottage at the empty east end of the island. We remember flying on to the beautiful island of Maui, staying at the native Maui Palms, where huge bowls were kept filled with guava and lilikoi juice in the lobby for its guests. Here we saw the ancient capital of Lahaina and dined under its astoundingly widespread banyans, and climbed up to the great Haleakala crater, snow there at the time. We remember Hilo, the "big" city on the biggest of the islands, Hawaii, with its snow capped 13,500-foot Maunakea volcano casting its brooding shadow over Hilo Bay and the hundreds of ancient and new lava flows radiating down all sides of the island from old dormant volcanoes and the still very active Kilawea, into whose boiling rock cauldron we peered in safety and comfort from the Volcano House on its flanks. Also we remember the gorgeous orchid plantations and the macademia nut farms and the Hawaiian cowboys in the little town of Kamuela, named after Samuel Parker, who started the great cattle ranch there.

We haven't forgotten our first visit to Kawai, the seemingly long flight from Honolulu to the Lihue airport, the tour through the big sugarcane processing plants near Port Arthur where the canestalks and leaves are pressed into bagasse and used as fuel for power generation, the stunning first look into the pastel-colored Waimea Canyon and our surprise at all the verdure and forests everywhere—until we learned this island had hundreds of inches of rainfall yearly and was probably one of the wettest spots on earth.

Yes, we fell in love with Hawaii at first sight and the cherished memories of its places and people have greatly enriched our lives.